THE KILLING OF CAMBODI
GENOCIDE AND THE UNM£

For my grandmother, Floris Tyner, and the children of Cambodia

The Killing of Cambodia: Geography, Genocide and the Unmaking of Space

JAMES A. TYNER
Kent State University, USA

Routledge
Taylor & Francis Group

LONDON AND NEW YORK

First published 2008 by Ashgate Publishing

2 Park Square, Milton Park, Abingdon, Oxon OX14 4RN
711 Third Avenue, New York, NY 10017, USA

Routledge is an imprint of the Taylor & Francis Group, an informa business

First issued in paperback 2016

British Library Cataloguing in Publication Data
Tyner, James A., 1966-
  The killing of Cambodia : geography, genocide and the
  unmaking of space
  1. Geopolitics - Cambodia 2. Genocide - Cambodia
  3. Cambodia - Politics and government - 20th century
  I. Title
  959.6'042

Library of Congress Cataloging in Publication Data
Tyner, James A., 1966-
  The killing of Cambodia : geography, genocide, and the unmaking of space / by James A. Tyner.
     p. cm.
  Includes bibliographical references and index.
  1. Geopolitics--Cambodia. 2. Genocide--Cambodia. 3. Cambodia--Politics and government--20th century. I. Title.

  JC319.T96 2008
  959.604'2--dc22

                                                          2007049173

ISBN 13: 978-0-7546-7096-4 (hbk)
ISBN 13: 978-1-138-25428-2 (pbk)

Contents

List of Plates

Acknowledgments

In *Abbey's Road*, Edward Abbey wrote, "The world is wide and beautiful. But almost everywhere, everywhere, the children are dying." These two lines have remained with me, not only during the completion of this book, but within my life. These fourteen words contain a sense of urgency, but also a spirit of optimism. The world is wide and beautiful; and I have been fortunate to have traveled throughout a great part of it. But injustices exist. Children are dying. They are dying from neglect and ignorance, exploitation and oppression. As academics, we have a responsibility both to the world and to its people.

First, I thank my commissioning editor at Ashgate, Valerie Rose. Val has been a tireless supporter of this project and I thank her deeply for her guidance, patience and vision. Without her, this book would not have been possible.

Over the years I have been fortunate to work and share ideas with a number of exceptional scholars; some were my advisors, others were my students, still others have been peers. All have been important. At the Document Center of Cambodia, I thank Youk Chhang and Sopheak Sim for their time and assistance. Closer to home, I thank the many colleagues who have contributed to my growth as an academic and, frankly, a person: Curt Roseman, Michael Dear, Stuart Aitken, Gary Peters, Keith Collins, Jan Monk, Laura Pulido, Colin Flint, Philip Kelley, Shannon O'Lear, Joseph Nevins, Larry Brown, Richard Wright, Adrian Bailey, Rachel Silvey, Scott Sheridan, Mandy Munro-Stasiuk, Rob Kruse, Josh Inwood, Gabe Popescu, Olaf Kuhlke, Donna Houston, Steve Butcher, Andrew Shears, and Stacey Wicker.

Heartfelt thanks are also extended to my family, Dr. Gerald Tyner, Dr. Judith Tyner, David Tyner, Karen Owens, Bill Owens. Their love, support, and guidance have been my constant companions. And, as always, I thank Bond (my puppy) and Jamaica (my cat), my late night-companions. My deepest thanks, though, must go to Belinda, my wife. She is quite literally the rock of our family. Without her, I doubt that I would be able to get through the day. And special thanks go to Jessica and Anica, my daughters. Aged five and six now, these girls continue to show me what is important. Life is about perspective. Thank you, Jessie and Anica, for keeping my perspective in focus.

During the completion of this book, my grandmother, Floris Tyner, passed away. She was but months away from reaching a century in age and it was impossible to not reflect on the forty-plus years I was able to talk and learn from my grandma as I wrote this manuscript. Grandparents are very special; they impart wisdom gained through life experiences. Grandma was no different. She was the heart of our family, the center of our universe. And through my writing, I thought also of the millions of people in Cambodia who died before their lives were complete. The Khmer Rouge attempted to kill a country and a people. And those that perpetrated that crime must be held accountable. And we, as academics, must not let the tragic details of that crime be forgotten. Consequently, this book carries two dedications. First, I dedicate this book to the memory of my grandmother and, second, through her memory I dedicate this to the children of Cambodia, many of whom were never able to know their grandparents.

Chapter 1

Imagining Genocide

And sometimes what he has to do for the state is to live, to work, to produce, to consume;
and sometimes what he has to do is to die.

Michel Foucault[1]

Cambodia's geography, we are taught, resembles a bowl, with a vast central plain surrounded by a series of mountains. In the southeast lie the Cardamom and Elephant Mountains and to the north and northeast, lie the Dangrek Mountains and Annam Highlands, respectively. According to Ian Brown, the "land and climate have helped to shape the Cambodian people and their way of life."[2] The region has a tropical climate, dominated by two monsoons. Beginning in late May or early June, the prevailing southwest winds bring the summer monsoon. Heavy rains and thunderstorms inundate Cambodia, causing the Mekong River to swell in size. This flooding, in turn, causes a tributary of the Mekong—the Tonlé Sap River—to reverse course. Consequently, Lake Tonlé Sap—that Great Lake which dominates the central Cambodia plain—expands four-fold in size. The enormous volumes of water, silt, and nutrients give rise to over 200 varieties of fish, and these form the major source of protein for the people of Cambodia.[3] In November, northeast winds dominate the region, bringing dry weather and cooler temperatures. The water-level of the Mekong drops, and the course of the Tonlé Sap River once again reverses its flow.

Over the millennia, these climatic conditions have influenced farming practices, certainly. But so too has the quotidian life—such as harvest festivals, religious ceremonies, wedding practices—of Cambodia's people been affected. Historically, the people of Cambodia—the Khmer—have been lowland wet-rice cultivators. The Khmer landscape has been that of villages and hamlets, with houses strung along roads, streams, and dykes.[4]

A vastly different geography was imagined (and thus attempted to be created) by Pol Pot. The secretive leader of the Communist Party of Kampuchea—the Khmer Rouge—Pol Pot viewed Cambodia (renamed Democratic Kampuchea under the Khmer Rouge) as a place of purity "amid the confusion of the present-day world,"

---

1 Michel Foucault, "Technologies of the Self," in *Ethics: Subjectivity and Truth, Essential Works of Foucault, 1954–1984*, Vol. 1, P. Rabinow (ed.) (New York: The New Press, 2000), 409.

2 Ian Brown, *Cambodia* (Herndon, VA: Stylus Publishing, 2000), 11.

3 John Tully, *A Short History of Cambodia: From Empire to Survival* (Crows Nest, Australia: Allen & Unwin, 2005), 3.

4 Michael D. Coe, *Angkor and the Khmer Civilization* (New York: Thames & Hudson, 2003), 40–42. See also Michael Vickery, *Society, Economics, and Politics in Pre-Angkor Cambodia* (Tokyo: Center for East Asian Cultural Studies for UNESCO, 1998).

"a precious model for humanity" whose revolutionary virtue exceeded that of all previous revolutionary states.[5] Pol Pot's geography contributed to the death of over two million of Cambodia's people—nearly one-third of the country's population—in less than four years.[6]

As a Geographer, I am most concerned with trying to understand the spatially-informed discursive foundations of the Khmer Rouge, of how the practitioners of genocide justified the material practices that led to the mass killings.[7] My starting point is that those people accused of genocide invariably show little or no remorse for their actions—irrespective of whether their actions are defined as genocide or not. In the aftermath of the Holocaust, for example, most Nazi leaders, including Adolf Eichmann and Heinrich Himmler, were unapologetic. They *believed* in their actions. Similar attitudes are found in other genocides, such as in Rwanda. And in Cambodia, Pol Pot exhibited no remorse at the time of his own death for the millions who perished under his reign. Other Khmer Rouge cadre, such as Nuon Chea and Ieng Sary, continue to proclaim their innocence of any wrong-doing. In all cases, these perpetrators of mass death were creating their own notion of social justice and, by extension, their own moral geographies.

Rather than adopting a psychological approach, of trying to get inside the minds of the "decision-makers" and their "followers," I am interested in the rationales provided, how these were justified and, in particular, how *geography* figured into these rationales. My intent is to repeatedly ask how certain *spatial* categorizations work, what enactments they are performing and what relations they are creating.[8]

---

5 Quoted in Philip Short, *Pol Pot: Anatomy of a Nightmare* (New York: Henry Holt and Company, 2004), 341.

6 The total number of deaths resultant from the genocide has been subject to intense debate. Demographers and anthropologists, activists and apologists, have all waded into the fray. Typical estimates range from a low of 1.5 million deaths to a high of approximately 3.3 million deaths. See, for example, Michael Vickery, "How Many Died in Pol Pot's Kampuchea?" *Bulletin of Concerned Asian Scholars*, 20(1988): 377–385; Ben Kiernan, "The Genocide in Cambodia, 1975–1979," *Bulletin of Concerned Asian Scholars*, 22(1990): 35–40; Patrick Heuveline, "'Between One and Three Million': Towards the Demographic Reconstruction of a Decade of Cambodian History (1970–1979)," *Population Studies*, 52(1998): 49–65; Craig Etcheson, "Did the Khmer Rouge Really Kill Three Million Cambodians?" *Phnom Penh Post*, April 30, 2000; Ben Kiernan, "The Demography of Genocide in Southeast Asia: The Death Tolls in Cambodia, 1975–1979, and East Timor, 1975–1980," *Critical Asian Studies*, 35(2003): 585–597; Damien de Walque, "Selective Mortality during the Khmer Rouge Period in Cambodia," *Population and Development Review*, 31(2005): 351–368; Craig Etcheson, *After the Killing Fields: Lessons from the Cambodian Genocide* (Lubbock: Texas Tech University Press, 2005); and Damien de Walque, "The Socio-Demographic Legacy of the Khmer Rouge Period in Cambodia," *Population Studies*, 60(2006): 223–231.

7 I should state up-front what this book is not. It is not an attempt to understand why the rank-and-file cadre carried out the violent practices (i.e., torture, execution, forced labor) that led to the millions of deaths. For an excellent discussion of this question, see Alexander L. Hinton, *Why Did They Kill? Cambodia in the Shadow of Genocide* (Berkeley: University of California Press, 2005).

8 Such a task is heavily indebted to that of Eve Sedgwick, *Epistemology of the Closet* (Berkeley: University of California Press, 1990), 27.

I am interested in the geopolitical discourses, the narratives and "spatial logics" that support, justify, and legitimate mass killings: In short, I want to understand the geographic imaginations constructed by state actors and how, in turn, these are used to justify "genocidal" practices.

How did the perpetrators of mass violence conceive of Cambodia's geography, and what type of spaces were they attempting to construct? Conceptually, these same concerns exist in the various attempts to understand genocides in other contexts and other places, such as Germany and Rwanda, Bosnia and Darfur. What type of society were the practitioners of genocide attempting to create? In the case of Cambodia, I identify one particular geographical imagination, one that directly influenced the machinations of violence. Here, I forward the idea of *anti-geographies*.

Many geographers have approached Geography from the stand-point that *geography* is literally about the writing of space. In the context of European colonialism, for example, Gearóid Ó Tuathail proposes that the practice of geography was "an active writing of the earth by an expanding, centralizing imperial state."[9] If this is so, then anti-geographies entail the erasure—the un-making, or un-writing—of space. This un-making, I argue, is foundational for the justification and legitimation of mass killings perpetrated by the Khmer Rouge.

My task, though, is immediately stalled by a linguistic problem. What is meant by genocide and mass killings? Did the killings—which are not denied by the perpetrators—constitute genocide? As will be clear momentarily, this question hinges, firstly, on how genocide is defined, and, secondly, whether the actions of certain individuals correspond with the agreed-upon definition of genocide. These are tremendously important, particularly in the effort to bring the Khmer leaders to justice, but these are also tangential to my larger project.

Since Raphael Lemkin coined the term "genocide," there have been numerous debates, discussions, and definitions forwarded. Rather than replicating these debates, I begin with the simple proposition that mass killing—genocide—is a political process. This statement also requires some explanation. Jenny Edkins makes a distinction between the "political" and "politics." The former, according to Edkins, is usually taken to be that sphere of social life commonly called *politics*, which would include elections, political parties, the state apparatus, and so on. Instead, Edkins suggests that the political "represents the moment of openness ... when a new social order is on the point of establishment, when its limits are being contested. Politics, in contrast, is what takes place once the new order is institutionalized."[10] Too often researchers focus on the consequences—the outcomes—of policies and practices, rather than tracing the genealogies of those actions. Edkins maintains that the political, however, is not a question of conflict between preexisting classes or other groupings whose interests can be presumed, but rather a question of the struggle to form new (and always precarious) coalitions of power.[11]

9 Gearóid Ó Tuathail, *Critical Geopolitics: The Politics of Writing Global Space* (Minneapolis: University of Minnesota Press, 1996), 2.

10 Jenny Edkins, *Poststructuralism and International Relations: Bringing the Political Back In* (Boulder, CO: Lynn Rienner Publishers, 1999), 2, 126.

11 Edkins, *Poststructuralism and International Relations*, 127–28.

Consequently, how something is recognized as *true*, *legitimate*, or *just*, is a political process. Just as academics and international lawyers continue to seek an "authentic" and "accurate" *representation* of genocide, so too have the practitioners of mass violence attempted to sanction their actions. This does lead to my overall concern: How did Pol Pot and the Khmer Rouge discursively legitimate and justify those material practices that led to mass killings? In anticipation of subsequent arguments, their justifications were *political*; they necessitated a particular representation of reality as they conceived of the real.

The Language of the Political

> *Political language ... is designed to make lies sound truthful and murder respectable ...*
>
> George Orwell[12]

For most people, language shapes their understandings of the world. Only rarely can there be direct observation of events and, accordingly, it is "language that evokes most of the political 'realities' people experience."[13] To this end, Murray Edelman argues that the political is linguistically constructed. Throughout *The Killing of Cambodia*, I employ a post-structural approach to understand the linguistic structuring of political reality.

There is no single definition of post-structuralism. Nor is there an agreed upon methodology or theory upon which to base one's research. Rather, the term refers to a collection of ideas—conceptual signposts—based on the writings Michel Foucault, Roland Barthes, Jacques Derrida, and Judith Butler, among others. For me, post-structuralism is an approach, an attitude, an awareness of possibilities.

Post-structuralists ask fundamental questions about language and subjectivity. In particular, post-structural approaches disrupt meanings and labels, categories and classification schemes; in other words, post-structuralism challenges terms that are assumed to be natural and unchanging. In so doing, post-structuralists propose that the distinctions we make are not necessarily given by the world around us, but are instead produced by the symbolizing systems we learn.[14] Furthermore, post-structuralism permanently troubles the notion that there can be a direct relationship between objects (i.e., genocide, justice) and the meanings they denote. Thus, there is no pre-interpreted reality (or explanation, for that matter) of genocide, nor is there an essential definition of genocide waiting to be discovered and agreed upon. Rather, the *meanings* of genocide are made discursively. Discourses of meaning are implicated in struggles for power and dominance between humans.[15] Genocide, even the act of identifying and defining genocide, is a *political process*.

---

12 George Orwell, *A Collection of Essays* (New York: Harcourt, 1981), 171.

13 Murray Edelman, *Political Language: Words that Succeed and Policies that Fail* (New York: Academic Press, 1977), 3.

14 Catherine Belsey, *Poststructuralism: A Very Short Introduction* (Oxford: Oxford University Press, 2002), 7.

15 Denis Cosgrove and Mona Domosh, "Author and Authority: Writing the New Cultural Geography," in *Place/Culture/Representation*, James Duncan and David Ley (eds) (New York: Routledge, 1993), 25–38; at 29.

Meaning, from a post-structural vantage point, is not guaranteed by a world external to it.[16] This implies that meanings are constructed. There is, consequently, no ontological referent to genocide. Indeed, this accounts for the cottage-industry that produces "new and improved" models and meanings of genocide. Instead, this term, in particular, is socially produced, politically contested, and temporarily fixed. When scholars or policy-makers use terms, such as genocide, they do so discursively.[17] However, the temporary fixing of meaning is never a neutral act; it involves both interests and questions of power.[18] Post-structuralism, consequently, through a focus on discourse, power, and knowledge, offers a revitalized framework from which to approach—but not finalize—the theoretical and empirical study of genocide.

A rudimentary understanding of semiotics is fundamental to all post-structuralism.[19] The Swiss linguist Ferdinand de Saussure proposed that meanings should not depend on reference to "real" objects or ideas. Rather, he proposed that meaning only resides in a *sign* and nowhere else. Saussure explained that a sign consists of two terms, a *signifier* (e.g., sound, written image, or graphic), and a *signified* (i.e., meaning). Barthes provides an example of a bouquet of roses. These roses (as signifier) may symbolize (signify) his passion for another person. However, we are not dealing with just two terms, a signifier (roses) and a signified (passion). Rather, a third term exists, this being the sign. A sign, therefore, is the associative total of the first two terms and derives meaning from being different from all other signs. Barthes emphasizes that the "roses" as signifier and "passion" as signified existed before uniting and forming the third object, the sign. Consequently, one should not confuse the roses as signifier and the roses sign: the signifier is empty, the sign is full, it is full of meaning.[20]

Furthermore, and this is the truly radical departure proposed by Saussure: the relationship between the signifier and the signified is purely arbitrary. The word "cat," for example, does not necessarily have to refer to the object which we have *learned* to identify as a cat. In other words, the signifier does not express the meaning, nor does the signified resemble the form or sound; to use a term appropriately—which, of course, may be politically contested—is to know what it means.[21] A classic example, is the traffic light as sign. The colors red, yellow, and green (in a North American context) each signify a particular action. Red, for example, "signals" drivers to stop, while green "permits" drivers to continue. These associations, however, are purely

---

16 Chris Weedon, *Feminist Practice and Poststructuralist Theory*, 2nd edition (Malden, MA: Blackwell Publishers, 1997), 171.

17 As political signs, terms such as "genocide" may be weakened through overuse.

18 Weedon, *Feminist Practice*, 171.

19 For more extensive discussions, see Vincent B. Leitch, *Cultural Criticism, Literary Theory, Poststructuralism* (New York: Columbia University Press, 1992); Madan Sarup, *An Introductory Guide to Post-Structuralism and Postmodernism*, 2nd edition (Athens: University of Georgia Press, 1993); Raman Selden and Peter Widdowson, *A Reader's Guide to Contemporary Literary Theory*, 3rd edition (Lexington: University of Kentucky Press, 1993); Belsey, *Poststructuralism*.

20 Roland Barthes, *Mythologies*, translated by Annette Lavers (New York: Hill and Wang, 1972 [1957]), 112.

21 Belsey, *Poststructuralism*, 11.

arbitrary. There is nothing inherent in the color red, for example, that denotes the action to stop. And indeed, in other contexts, the color red may symbolize other actions or even emotions (e.g., love, anger, or danger).

Saussure and other "structuralists" attempted to "fix" meanings within language systems. The problem with this perspective, as Weedon explains, is that it does not account for the plurality of meaning or for changes—whether historical or geographical—in meaning.[22] In contrast, post-structuralists have demonstrated the malleability of signs. Raman Seldon and Peter Widdowson, for example, illustrate that the dictionary confirms the relentless deferment of meaning, as opposed to the unity of any given sign:

> Not only do we find for every signifier several signifieds (a 'crib' signifies a manger, a child's bed, a hut, a job, a mine-shaft linign, a plagiarism, a literal translation, discarded cards at cribbage), but each of these signifieds becomes yet another signifer which can be traced in the dictionary with its own array of signifieds ('bed' signifies a place for sleeping, a garden plot, a layer of oysters, a channel of a river, a stratum). The process continues interminably, as the signifiers lead a chameleon-like existence, changing their colors with each new context.[23]

The indeterminancy and contingency of language is further—and aptly—illustrated in Derrida's well-known discussion of the work *pharmakon*. A Greek word, it can mean either "poison" or "cure," depending on the context of translation.[24] Poststructuralists, consequently, argue that no particular signifier can refer to any particular signified; indeed, every signified functions in turn as a signifier in an endless play of signification.[25] The meaning of "genocide," as a sign, is thus open to multiple interpretations. As a corollary, certain actions (i.e., the killing of people) may or may not be defined and interpreted as a genocide. This, then, accounts for the inactions of governments throughout the twentieth-century when confronted with mass killings. To understand the debates surrounding genocide in these terms, therefore, "not as struggles between bias and accuracy, nor as repeated and ever more accurate revisions based on improved information" but rather as "the ongoing contention between competing political and philosophical perspectives" permits us "to see the repeated overturning of interpretations not as a futile and absurd quest for the truth but as a social and political struggle over the production of meaning."[26]

---

22 Weedon, *Feminist Practice*, 24.

23 Seldon and Widdowson, *Contemporary Literary Theory*, 126–27.

24 The seemingly contradictory "nature" of *pharmakon* as either "poison" or "cure" foreshadows another important concept, namely that constructed meanings, or discourses, are neither inherently good or bad. Such evaluative claims must necessarily be context-specific. It is for this reason that perpetrators of genocide have been able to justify (in their minds, at least) the killing of massive numbers of people.

25 Sarup, *Introductory Guide to Post-Structuralism*, 41.

26 Ellen Somekawa and Elizabeth A. Smith, "Theorizing the Writing of History or, 'I Can't Think Why it Should be so Dull, for a Great Deal of it Must be Invention,'" *Journal of Social History*, 22(1988): 149–162; at 155.

The political is constructed linguistically. However, language always exists in historically and geographically specific *discourses* which inhere in social institutions and practices. It is to this concept—discourse—I now turn.

Discourse and the Power/Knowledge Nexus

Knowledge about any given object, and especially the production of that knowledge, is crucial to an understanding of political practices, including genocide. Here, I draw liberally on the ideas forwarded by Michel Foucault. Throughout his many writings, Foucault provided a critique of the way modern societies control and discipline their populations by sanctioning the knowledge claims and practices of the human sciences.[27] This is particularly salient when one considers genocide, in that knowledge claims were widely employed to justify the killing of people. As scholars have exhaustively documented, the murder of Jews, for example, in the Nazi death camps was approved by members of the medical establishment and other scientists.

A Foucauldian approach is concerned with how knowledge is produced and subsequently deployed. We may assert, for example, that the construction and forwarding of particular policies that "justify" and "legitimate" mass killings is predicated on particular "knowledges;" these knowledges, furthermore, are derived from the efforts of academics, activists, policy-makers, and so on. These knowledges, likewise, are produced via research, personal beliefs or observation, and various forms of "data collection."

Given that knowledge is neither neutral nor value-free, but rather the outcome of political processes, it follows that knowledge is intricately associated with social relations. Moreover, given that social relations are not necessarily asymmetrical, one is left with the proposition that "power" assumes a pivotal role in knowledge production. And in fact, Foucault forwards the idea that power and knowledge directly imply one another; that there is no power relation without the correlative constitution of a field of knowledge, nor any knowledge that does not presuppose and constitute at the same time power relations.[28]

Power, though, should not be conceived *a priori* as a negative force that one may use to coerce another. From a Foucauldian perspective, power differs from traditional accounts: Power is exercised rather than possessed. Furthermore, power is conceived as being inherent within social relations. Power circulates; power is exercised. This proposition, of power as being exercised, is significant in that it directs our attention from focusing on a dominant class or institution, and toward the relations between different groups of people. Power and the production of knowledge are not simply the result of an oppressive capitalist system or its apparatuses. On this count, Foucault is clear: power is not the privileged domain of a dominant class; authorities do not have a monopoly on the exercise of power, or on the production of knowledge.[29] In his

---

27 Sarup, *Introductory Guide to Post-Structuralism*, 72.

28 Michel Foucault, *Discipline and Punish: The Birth of the Prison*, translated by A. Sheridan Smith (New York: Vintage Books, 1979), 27.

29 Foucault, *Discipline and Punish*, 26.

History of Sexuality, Foucault explains that power is not something that is acquired, seized, or shared; relations of power are not in a position of exteriority with respect to other types of relationships (e.g., economic processes); there is no binary and all-encompassing opposition between rulers and ruled at the root of power relations; there is not power that is exercised without a series of aims and objectives; and there is no exercise of power without a correlative resistance.[30]

Researchers, accordingly, should direct their attention to the exercise of power in the production of knowledge vis-à-vis genocide. Who has the authority to *define* certain groups as "unworthy of life"? Who has the authority to *justify* and *legitimate* the mass killing of these people? How are these violent and inhumane policies and practices *explained* to the public? In short, a focus on the power/knowledge nexus leads to a greater engagement with discourses.

Discourse, for our present discussion, refers not to the standard dictionary definition of a conversation or written expression, but rather to *disciplines of knowledge*. In *The Archaeology of Knowledge*, Foucault's principal text in which he outlines his coordinates of discourse and discursive formations, discourse is derived from statements which have four attributes.[31] First, a statement must have a material existence. By this, Foucault simply means that statements must appear: they must be articulated. The form of statements, however, is open. Political speeches, policies, laws and regulations, most assuredly, are statements. So too, however, are graphs, charts, maps, censuses, history books, personal identification cards, and so on.[32] Second, and relatedly, statements must have a substance; they are manifest in particular places and at specific times. When these contexts change, so too does the discourse. Third, statements do not have as their correlate an individual or a particular object that is designated by this or that word. This is crucial, for it directs attention to the fact that there is no "true" referent that we are attempting to describe. Statements do refer to objects, but these do not derive in any sense from a particular state of things, but stem from the statement itself.[33] Again, when politicians or academics refer to genocide, they are not referring to a fixed object, for there is no "true" or "essential" genocide. A genocide only *becomes* a genocide when relevant state actors assert—through political statements—and agree that a situation of mass killings is a genocide. Is this a tautological argument? Yes, it most certainly is. But how else can one account for the political vicissitudes of the twentieth-century that have shrouded discussions of genocide? What else accounts for the inaction of the United Nations during the spring and summer of 1994 as hundreds of thousands of people were slaughtered in Rwanda? Why else has the international community

30 Michel Foucault, *The History of Sexuality, Volume I: An Introduction*, translated by R. Hurley (New York: Vintage Books, 1990), 94–95.

31 Michel Foucault, *The Archaeology of Knowledge & the Discourse on Language*, translated by A. Sheridan Smith (London: Pantheon Books, 1972), 91–99.

32 As discussed later, the Khmer Rouge made extensive use of slogans and songs to disseminate political statements among the people.

33 Gilles Deleuze, *Foucault*, translated by S. Hand (Minneapolis: University of Minnesota Press, 1988), 7–8.

(at the time of writing) disagreed as to whether the on-going violence in Darfur constitutes a genocide?

Foucault forwards a number of principles that guide his understanding of discourses. To begin with, a discourse must be treated as a discontinuous activity, its different manifestations sometimes coming together, but just as easily excluding each other. In other words, there is often considerable agreement among the various discourses of genocide forwarded by scholars, politicians, or even those standing trial for the crime of genocide. However, despite these similarities, there are also remarkable divergences, which lead to different interpretations and evaluations. Foucault refers to this as the principle of discontinuity. Also, Foucault proposes a principle a "rarity," arguing that "everything is never said."[34] All knowledges are partial and, consequently, discourses are selective. However, the absences, or silences, are not to be automatically thought of as an indication of repression. This is important when considering government-sanctioned discourses. We cannot presume—although we may, through empirical work, argue—that discursive practices conceal a hidden agenda. Further, Foucault forwards a principle of specificity: a particular discourse cannot be resolved by a prior system of significations. In other words, we should not imagine that the world presents us with a legible face, leaving us merely to decipher it.[35] Genocide is not "waiting to be discovered;" it must be agreed upon. Lastly, discourses, for Foucault, are limited. They restrict what is or is not to be included.

A Foucaldian approach to discourse, in sum, is to view discourses not as groups of signs (signifying elements referring to contents or representations) but as *practices that systematically form the objects of which they speak*.[36] The knowledge that enables us to answer questions, to make policy, to make judgements pertaining to a particular actions is informed—indeed produced—by the discursive practices constituting and demarcating the field. Consequently, there can never be a singular discourse of genocide, but rather multiple, contradictory, and contested discourses of genocide.

A post-structural approach to genocide, as such, identifies two different tasks which, in general, have defined the field of "genocide studies." On the one hand, there is the task of defining and arguing whether a particular event, such as the mass killing of people, should be considered a genocide. This task has been largely assumed by legal scholars who attempt to prosecute the perpetrators accused of genocide. On the other hand, there is the task of understanding the motivations, the intentions, the knowledge-claims forwarded by the perpetrators.[37] Social scientists, and especially historians (but also sociologists and anthropologists), have mostly engaged in this

---

34 Foucault, *Archaeology of Knowledge*, 118.

35 Foucault, *Archaeology of Knowledge*, 229.

36 Foucault, *Archaeology of Knowledge*, 49.

37 I may, at this point, be accused of side stepping the moral question of right and wrong implied by the term *genocide*. This is, in fact, a question I explore in the final chapter, namely, the implications for social justice. Suffice it to say, I believe social justice is likewise discursive, resultant from a political process. There can be, in other words, no "universal" human rights. Rather than promoting a nihilistic position, I will argue that this stance is, in fact, more liberating. Consider: Human rights are not pre-ordained, handed down by some omniscient being. Rather, someone, or some group of people, must decide what constitutes a

task. Geographers, curiously, have been mostly silent on the subject of genocide. This is surprising, given that key *spatial* concepts may provide important insights into the intentions of the perpetrators of genocide.

Geographical Imaginations

The writing of our geographies is a process of creating and inscribing meanings about our places and spaces.

Denis Cosgrove and Mona Domosh[38]

The struggle over geography is also a conflict between competing images and imaginings ...

Gearóid Ó Tuathail[39]

Peter Taylor recently suggested that "God invented war to teach Americans geography."[40] Of course, he is being deliberately provocative, his statement designed to induce engagement. Nevertheless, there is a rather large kernel of "truth" contained within part of his statement, namely that war does facilitate geographic interest—though not necessarily understanding. Susan Schulten notes that following certain pivotal moments during the Second World War—the German invasion of Poland, the bombing of Pearl Harbor, and the assault on Normandy—Americans bought in a matter of hours what in peacetime would have been a year's supply of maps and atlases. She continues that this nationwide attention to maps brought the farthest reaches of the war into everyday conversation, and demonstrated the powerful relationship between war and geography.[41]

Raising the connection between war and geography at this point is not to suggest that Geographers are Mars' handmaidens. Rather, it is to emphasize the importance of geographic knowledges. Geography, as Schulten explains, is a way of distinguishing "here" from "there," without which little sense can be made of human experience. However, and this relates directly to the urgency and immediacy associated with warfare, wars have deliberate and explicit effects on the *nature* of geographic knowledge and of spatial representation. For example, Schulten elaborates that Americans pored over new maps—namely, Goode's "homolosine projection"—that highlighted America's *proximity* to Europe and Asia, shaking the nation's well-developed sense of isolation.[42] Maps, therefore, were just one medium

---

universal human right. *Would you want Pol Pot, Joseph Stalin, or Adolf Hitler to make that decision?*

38 Cosgrove and Domosh, "Author and Authority," 27.

39 Ó Tuathail, *Critical Geopolitics*, 14.

40 Peter J. Taylor, "God Invented War to Teach Americans Geography," *Political Geography* 23(2004): 487–492.

41 Susan Schulten, *The Geographical Imagination in America, 1880–1950* (Chicago: University of Chicago Press, 2001), 1–2.

42 Schulten, *Geographical Imagination*, 2–3.

(among many) that re-defined America's geographic knowledge—imagination—of the world.[43]

But what exactly is meant by "geographic knowledge"? In common parlance, the phrase often refers to some understanding, some knowledge, about a particular place. Consequently, geographical knowledge may be understood as that information purported to explain, describe, and interpret the distributions and characteristics of people and places. This is a fairly standard approach to Geography: the writing of peoples and places. Conversely, geographical knowledge may also encompass a normative dimension in that it prescribes where people are to be located. David Harvey explains that states, for example, may institute normative programs for the production of new geographical configurations and in so doing become major sites for orchestrating the production of space, the definition of territoriality, the geographical distribution of population, economic activity, social services, wealth and well-being.[44] In either case (the former being more descriptive, the latter being more normative), geographic knowledges are representations.

Edward Said referred to these knowledges as *imaginative geographies*.[45] These "ways of seeing," according to Said, "legitimate a vocabulary, a representative discourse peculiar to the understanding of places that becomes *the* way in which a place is known."[46] Imaginative geographies are, in effect, geopolitical discourses—metageographies—that provide the foundation of foreign policy. As explained by Lewis and Wigen, metageographies may be conceived as "spatial structures"—geopolitical visions—through which people order their knowledge of the world. They maintain that diplomats, politicians, and military strategists employ metageographical frameworks no less than do scholars and journalists. Lewis and Wigen note, for example, that during the Cold War Americans relied heavily on a tripartite classification scheme to give order to the map—the familiar First, Second, and Third World divisions. More recently, new metageographies have emerged, such as US President George W. Bush's "Axis of Evil." Lewis and Wigen warn that such geographical constructs do not simply lead to faulty understandings of human societies but instead constitute ideological structures.[47] Derek Gregory concurs, noting that imaginative geographies, or metageographies, involve a politics of space. He asserts, "Who claims the power to represent: to imagine geography like this rather than like that? The process of articulation is … also a process of valorization."[48] In short, geographic representations undergird political practices. Consequently, as Lewis

---

43 For more in-depth discussions on the relationship between war and the development of (Anglo-American) geographical thought, see Trevor J. Barnes and Matthew Farish, "Between Regions: Science, Militarism, and American Geography from World War to Cold War," *Annals of the Association of American Geographers* 96(2006): 807–826 and Felix Driver, *Geography Militant: Cultures of Exploration and Empire* (Oxford: Blackwell, 2001).

44 David Harvey, *Spaces of Capital: Towards a Critical Geography* (New York: Routledge, 2001), 213.

45 Edward Said, *Orientalism* (New York: Vintage Books, 1979).

46 Bill Ashcroft and Pal Ahluwalia, *Edward Said* (New York: Routledge, 1999), 61.

47 Martin Lewis and Kären Wigen, *The Myth of Continents: A Critique of Metageography* (Berkeley: University of California Press, 1997)

48 Derek Gregory, "The Lightning of Possible Storms," *Antipode* 36(2004): 798–808.

and Wigen conclude, when used by those who wield political power, the outcomes can be truly tragic; indeed, metageographies may be used as tools of genocide.[49]

In short, there is no *presentation*, only *re-presentation*.[50] However, it was not until the early 1990s that Geographers and other social scientists began to seriously question how spaces and places, people and events, are represented.[51] Traditionally, James Duncan and David Ley explain, Geographers participated in "descriptive fieldwork" whereupon the objective of research was to observe and record—through proper training—the world and its inhabitants. Such an approach was supplemented, or in some circles, supplanted, by a more positivist orientation. Associated with the "quantitative" revolution of the 1950s and 1960s, Geographers attempted to describe and understand the world through the use of various statistical measures. Undergoing an ontological and epistemological revolution—as opposed to a methodological revolution—Geography was soon dominated by practitioners who sought to adopt the principles and practices of scientific investigation.[52]

Positivist science, broadly defined, consists of several characteristic features: the collection of data through observation and measurement of things that are "known" to exist and can be directly experienced; the development of generalizations and deduced laws that can only follow on the basis of repeated observations and the testing of hypotheses about causal relations that exist between phenomena; to combining of accepted generalizations and hypotheses into theories and laws that explain how the world works; and the assumption that theories can never be completed validated (i.e., verified) in the sense of proved absolutely correct, but can be provisionally accepted until contrary evidence or data are collected.[53]

Both descriptive fieldwork and positivist approaches are considered to be *mimetic*, meaning that each approach attempts to provide an accurate representation of reality. Many of the foundational assumptions underlying these approaches, however, were critiqued and challenged by Geographers during the 1980s and 1990s, culminating in what some referred to as a "crisis of representation." Specifically, some Geographers (and other social scientists) questioned the long-held beliefs of an independently existing reality and the mimetic theory of representation; others questioned the ability to accurately record and interpret reality, however defined; and still others rejected the search for "truth" and the totalizing ambitions of modern science, i.e., the search

---

49 Lewis and Wigen, *The Myth of Continents*, ix–xiii, 8. See also James A. Tyner, *America's Strategy in Southeast Asia: From the Cold War to the Terror War* (Boulder, CO: Rowman & Littlefield, 2007).

50 This theme is well-developed in the writings of Michel Foucault, Roland Barthes, Edward Said, and those scholars informed by post-structuralism.

51 Geography, certainly, was not alone. According to Roxanne Doty, the question of representation has historically been excluded from the academic study of international relations. Roxanne Lynn Doty, *Imperial Encounters: The Politics of Representation in North-South Relations* (Minneapolis: University of Minnesota Press, 1996), 3.

52 Duncan and Ley, "Introduction," 2–3.

53 Phil Hubbard, Rob Kitchin, Brendan Bartley, and Duncan Fuller, *Thinking Geographically: Space, Theory and Contemporary Human Geography* (New York: Continuum, 2002), 29; see also Ron Johnston, *Philosophy and Human Geography: An Introduction to Contemporary Approaches*, 2nd edition (London: Edward Arnold, 1986), 11–54.

for universal laws.[54] Derek Gregory is blunt in his assessment: Human geographers must work with social theory because we have little choice. "Empiricism," for Gregory, "is not an option" because the "facts" do not (and never will) "speak" for themselves, no matter how closely we listen.[55]

According to Denis Cosgrove and Mona Domosh, the "scientific way of knowing is no longer regarded as a privileged discourse linking us to truth but rather one discourse among many, which constructs both the object of its enquiry and the modes of studying and representing that object."[56] Doty explains, for example, that "International relations are inextricably bound up with discursive practices that put into circulation representations that are taken as 'truth.'"[57] This expands upon Foucault's articulation of truth: Truth, for Foucault, is centered on the form of scientific discourse and the institutions which produce it. Truth, accordingly, is "subject to constant economic and political incitement (the demand for truth, as much for economic production as for political power); it is the object, under diverse forms, of immense diffusion and consumption (circulating through apparatuses of education and information whose extent is relatively broad in the social body, not withstanding certain strict limitations; it is produced and transmitted under the control, dominant if not exclusive, of a few great political and economic apparatuses (university, army, writing, media); lastly, it is the issue of a whole political debate and social confrontation."[58]

The "crisis of representation" is vitally important to contemporary discussions of genocide and mass violence, for its highlights the processes involved in defining what constitutes genocide. Numerous scholars, for example, have argued that one defining feature of genocide is the deliberate killing of civilians. Such a distinction serves ostensibly to separate genocidal practices from those of warfare. Shaw concludes that the "difference between war and genocide is not the destructive character of the action, the violent modality, or the typical actor." Rather, the "difference lies in the construction of civilian groups as enemies, not only in a social or political but also in a military sense, to be destroyed." According to Shaw, genocide, unlike war, "constructs unarmed civilian populations as the objects, in their own right, of the types of armed violence normally applied only to armed enemies."[59]

Supposedly, this distinction removes other violent acts—the aerial bombings of Hamburg and Dresden, the nuclear destruction of Hiroshima and Nagasaki, the environmental warfare waged over Vietnam—from the realm of "genocide." Shaw, therefore, is able to argue that "although mass slaughter of civilians—for example at Dresden or Hiroshima—was evil and illegitimate, it could still be understood at least

---

54 Duncan and Ley, "Introduction," 2–3.

55 Derek Gregory, "Intervention in the Historical Geography of Modernity: Social Theory, Spatiality and the Politics of Representation," in *Place/Culture/Representation*, James Duncan and David Ley (eds) (New York: Routledge, 1993), 272–313, at 275.

56 Cosgrove and Domosh, "Author and Authority," 28.

57 Doty, *Imperial Encounters*, 5.

58 Michel Foucault, "Truth and power," in C. Gordon (ed.) *Power/Knowledge: Selected Interviews and Other Writings, 1972–1977*, translated by C. Gordon, L. Marshall, J. Mepham, and K. Soper (New York: Pantheon Books, 1980), 109–33; at 131–32.

59 Martin Shaw, *What is Genocide?* (Malden, MA: Polity, 2007), 111–112.

partially under the rubric of war. Civilians were part of the enemy; but no civilian group was an enemy distinct from the enemy state."[60] Chalk and Jonassohn likewise exclude civilian victims of aerial bombardment in belligerent states from *their* definition of genocide. Chalk and Jonassohn maintain that in this age of total war belligerent states make all enemy-occupied territory part of the theater of operations regardless of the presence of civilians. Civilians are regarded as combatants so long as their governments control the cities in which they reside.[61] Such a stance contrasts with that of Leo Kuper who has argued that the bombing of civilian enemy populations does constitute genocide.

Herein lies the problem of representation. Genocides are defined externally. At no point throughout the twentieth-century since the term "genocide" was coined has any one identified their own policies or practices as genocide. Neither Pol Pot nor Slobidan Milosivic, for instance, characterized their policies—which led directly to the deaths of many hundreds of thousands of people—as genocide. Charges of genocide are always raised by other people, other institutions, other governments, than those perpetrating violent acts. What counts as genocide, therefore, as opposed to some other "legitimate" form of mass killing, is a political process. It is not surprising that many American scholars would not consider the annihilation of Native Americans as a genocide; it is also not surprising that such scholars would not consider the brutality associated with the enslavement of Africans in the United States genocidal. And it is not surprising that many scholars would not consider the loss of life associated with the use of napalm and cluster bombs in Vietnam and Cambodia by the US military as genocidal acts. Following Cosgrove and Domosh, judgements that we make—including those judgements directed toward genocide— are not based on any concept of empirical truth but rather lie in a realm of moral discourse, a central element in the struggle for meaning.[62]

Geographic knowledges are not mimetic representations of reality or truth. Rather, they are the product of the political; they are produced. How then do we interpret the underlying geographical imaginations of the mass killing of people? I begin with the premise that mass killings—practices that become labeled as genocide—are neither natural nor inevitable. Rather, these result from the deliberate forwarding of actions that lead to widespread death and destruction. Such practices may be direct and explicit in their murderous intent, such as the gassing of Jews and other persons during the Holocaust, as well as other forms of executions and bombings. Practices may also indirectly contribute to mass killings, such as starvation, exposure, or disease resulting from forced relocation/forced labor, or from the intentional confiscation, destruction, or blockade of the necessities of life.[63] Given such a stance, I concur

---

60 Shaw, *What is Genocide*, 112. It should be noted, further, that not all scholars would consider the bombing of Dresden or Hiroshima as evil or illegitimate. Such a statement reflects yet another contested representation.

61 Frank Chalk and Kurt Jonassohn, *The History and Sociology of Genocide: Analyses and Case Studies* (New Haven, CT: Yale University Press, 1990).

62 Cosgrove and Domosh, "Author and Authority," 29.

63 Benjamin A. Valentino, *Final Solutions: Mass Killings and Genocide in the 20th Century* (Ithaca, NY: Cornell University Press, 2004), 10.

with Valentino who argues that the effort to understand mass killing should begin with an examination of the capabilities, interests, ideas, and strategies of groups (i.e., the Khmer Rouge) and individuals (i.e., Pol Pot, Khieu Samphan) in positions of political and military power and not with factors that predispose societies to produce such leaders.[64] What is required, following Valentino, is an attempt to understand the "strategic logic" of mass killing.

However, we must also recognize that any strategic logic is inherently a *spatial logic*. In other words, a particular and normative geographic knowledge undergirds the ideas and strategies of groups and individuals; a specific geographic vision is sought for society, a national myth is to be constructed. Such a geographic imagination thus addresses many key aspects of mass killing: Who is to be included in, or excluded from, the desired society? What is the most effective way of achieving the desired society (i.e., relocation or the elimination through death of people)? How are social relations to be arranged in the new society?

Our task is to uncover the "battle for truth," those discourses—those spatial representations—that promote and facilitate, justify and legitimate, the *sanctioned* killing of people. The task of analysis, therefore, is "not to reveal essential truths that have been obscured, but rather to examine *how* certain representations underlie the production of knowledge and identities and how these representations make various courses of action possible."[65] In *The Killing of Cambodia*, consequently, I attempt to chart out those geographical imaginations that justified, for the Khmer Rouge, the death of a people and a place.

---

64 Valentino, *Final Solutions*, 64.

65 Doty, *Imperial Encounters*, 5.

Chapter 2

Irruptions and Disruptions

The aim of our revolutionary struggle is to establish state power within the grasp of the worker-peasants, and to abolish all oppressive state power.[1]

Born in 1926, Keo Meas devoted his entire life to the Cambodian Communist movement. And like many revolutionaries, his initial foray into Communism came via his desire to liberate his homeland from colonial domination. Soon after the Second World War, at the age of 15, Meas dropped out of his teacher training courses at the Phnom Penh Teacher Training College to join a Khmer Viet Minh group in Svay Rieng province. In 1950 he was one of only 21 Cambodian members of Ho Chi Minh's Indochinese Communist Party (ICP). Having proven himself, he was appointed Commissar of the Action Committee for Phnom Penh, and in 1952 he traveled to Beijing. He was the first Khmer to meet Chairman Mao Zedong.

Keo Meas began to see himself as the future leader of the yet-to-be-established Cambodian Communist Party. He was diligent in his efforts, disseminating propaganda via the "Voice of Free Cambodia" radio station—of which a soft-spoken and charismatic man named Saloth Sar (later known as Pol Pot) wrote commentaries.[2]

In 1954 Keo Meas assumed the position of secretary of the Phnom Penh Committee of the Khmer People's Revolutionary Party (KPRP). Working in the capital, Meas continued to serve the Communist movement. He participated in national elections as a member of the Pracheachon ("People's Group"), a political party which he in fact helped establish. Although the Pracheachon's statutes contained no reference to communism, it was widely known as the legal Communist political party. It was through the Pracheachon that many Khmer Communists attempted to operate within mainstream Cambodian politics.[3]

In 1960 Keo Meas was elected to the Central Committee of the newly-formed (and clandestine) Workers' Party of Kampuchea (WPK). And although he lost this position in 1963, he remained in good standing with the Communist movement. In the early 1970s, as the Communist movement approached victory,

---

1 "Decisions of the Central Committee on a Variety of Questions," in *Pol Pot Plans the Future: Confidential Leadership Documents from Democratic Kampuchea, 1976–1977*, translated and edited by David P. Chandler, Ben Kiernan, and Chanthou Boua (New Haven, CT: Yale University Southeast Asia Studies, 1988), 6.

2 Philip Short, *Pol Pot: Anatomy of a Nightmare* (New York: Henry Holt and Company, 2004), 99–100.

3 Short, *Pol Pot*, 107.

Keo Meas served as ambassador to China for the Royal Government of the National Union of Kampuchea.

By all accounts, Keo Meas was a dedicated believer and leader of the Communist revolution in Cambodia. But on September 20, 1976 he was arrested and taken to Tuol Sleng, the Khmer Rouge's detention center located in Phnom Penh. There, Keo Meas was interrogated and tortured for over a month before his execution. Transcripts of his interrogation totaled 96 pages in his own handwriting. Among these documents are two lengthy confessions, one of which was entitled "1951 or 1960."[4]

Keo Meas' ordeal is particularly salient in a geneaology of the Khmer Rouge. Michel Foucault warned of the search for orginary moment. As the Orwellian ordeal of Keo Meas demonstrates, such a quest marks the difference between life and death. His plight also supports my positioning of the geneaology of the Khmer Rouge as both a social revolution and a resistance movement.

At issue is the founding date of the Cambodian Communist movement. When Keo Meas was arrested, there had been no public announcement of the party's existence. And yet, two magazine stories appeared in 1976, buttressing Keo Meas's detention. The September 1976 issue of the Communist Party of Kampuchea's (CPK) youth magazine, *Yuvechon nung Yuvanearei padevat* ("Revolutionary Youths and Maidens") opened with a 16 page article celebrating the *25th anniversary* of the party's founding. The article began: "From the moment of its creation on September 30, 1951, the Communist Party of Kampuchea led everyone, including revolutionary Cambodian youths, in the struggle against French imperialism."[5] The following month, the party's official journal, *Tung Padevat* ("Revolutionary Flag"), opened with a 32 page article, commemorating the party's *16th anniversary*. Most likely written by Pol Pot, the article explained: "Last year we informed people ... that our Party was 24 years-old ... But now we celebrate the 16th anniversary of the party, because we are making a new numeration. What rationale is there for this? The revolutionary organization had decided that from now on we must arrange the history of the party into something clean and perfect, in line with our policies of independence and self-mastery."[6] The making of a new Cambodia "clean and perfect" required an eradication of all that was deemed inconsistent with CPK goals.[7]

The death of Keo Meas provides a grim lesson in our understanding of the growth of the Khmer Rouge. It provides insight into the mind-set and the motivations of key

---

4 David P. Chandler, "Revising the Past in Democratic Kampuchea: When Was the Birthday of the Party?" *Pacific Affairs* 56(1983): 288–300; at 288; see also David P. Chandler, *Voices from S-21: Terror and History in Pol Pot's Secret Prison* (Berkeley: University of California Press, 1999), 58–60.

5 Chandler, "Revising the Past," 289.

6 Chandler, "Revising the Past," 289.

7 According to Chandler, it is likely that Pol Pot, along with Nuon Chea (Brother Number Two), wrote both articles in an attempt to draw "1951" factions into the open, where they could then be eliminated. At the time, those who favored a 1951 date were presumed to be in alliance with the Vietnamese Communists, and thus traitorous. This episode would spur a major purge of CPK members throughout 1976 and 1977. See Chandler, *Voices from S-21*, 60.

leaders of the Communist movement in Cambodia. It also serves as a backdrop for my subsequent discussion of the CPK's revolution, its relationship with Vietnam, and its justification for mass killings.

There are many accounts—re-presentations—of the "origins" of communism and the subsequent genocide in Cambodia.[8] As studies of Cambodia and other genocides indicate, however, historical placement is often difficult. Zygmunt Bauman, for example, explains that initially, he—along with many other sociologists—believed (by default rather than by deliberation) that the Holocaust "was an interruption in the normal flow of history, a cancerous growth on the body of civilized society, a momentary madness among sanity."[9] Considered reflection, however, forced Bauman to reevaluate how the Holocaust is re-presented in our classrooms and our research. Likewise, we must be deliberate in our reconstructions of communism and of other resistance movements in Cambodia that preceded the horrific killings between 1975 and 1979.

Foucault warned against the search for origins and, in the process, questioned the initial quest for, or discoveries of, historical continuities.[10] Through his writings Foucault suggested that we temporarily suspend certain assumptions. We must question, for example, the "principle of coherence," the idea that there exists ready-made syntheses that are normally accepted before any examination. From a Foucauldian standpoint, there can be no *assumed* constants, no essences, no immobile forms of uninterrupted continuities structuring the past.[11] Nor should we be seduced by abstract and (often) arbitrary periodizations (e.g., decades or presidential terms). We should, moreover, guard against the assignment of an irruption of a real event; we should question the attempt, even, to assign a singular date on which a particular event, or movement, began.[12] We should also resist the

---

8 Apart from those listed earlier, see also Ben Kiernan, "Origins of Khmer Communism," *Southeast Asian Affairs*, 1981: 161–180; Serge Thion, "The Cambodian Idea of Revolution," in *Revolution and its Aftermath in Kampuchea: Eight Essays*, D.P. Chandler and B. Kiernan (eds) (New Haven, CT: Yale University Southeast Asian Studies, 1983), 10–33; David P. Chandler, "Seeing Red: Perceptions of Cambodian History in Democratic Kampuchea," in *Revolution and its Aftermath in Kampuchea: Eight Essays*, D.P. Chandler and B. Kiernan (eds) (New Haven, CT: Yale University Southeast Asian Studies, 1983), 34–56; David P. Chandler, *Brother Number One: A Political Biography of Pol Pot*, revised edition (Chiang Mai, Thailand: Silkworm Press, 2000); David P. Chandler, *A History of Cambodia*, 3rd edition (Boulder, CO: Westview Press, 2000); Ben Kiernan, *How Pol Pot Came to Power: A History of Communism in Kampuchea, 1930–1975* (London: Verso, 1985); Ben Kiernan, *The Pol Pot Regime: Politics, Race and Genocide in Cambodia under the Khmer Rouge, 1975–1979* (New Haven, CT: Yale University Press, 1996).

9 Zygmunt Bauman, *Modernity and the Holocaust* (Ithaca, NY: Cornell University Press, 1989), viii.

10 Michel Foucault, *The Archaeology of Knowledge and the Discourse on Language*, translated by A. Sheridan Smith (New York: Pantheon Books, 1972).

11 Madan Sarup, *An Introductory Guide to Post-Structuralism and Postmodernism*, 2nd edition (Athens: University of Georgia Press, 1993), 59.

12 Ben Kiernan, for example, identifies a 24 year-old indentured laborer named Ben Krahom as "the first Khmer known to have become involved in communist activities" in Cambodia. See Kiernan, "Origins of Khmer Communism," 161–162.

temptation to identify a straight-forward cause-and-effect relationship, or a defining "moment" (the elusive "smoking gun") that is proximate to the killings.[13] Such a task is impossible, for, as Foucault identifies, "one is led inevitably, through the naïveté of chronologies, towards an ever-receding point that is never itself present in any history." Consequently, "all beginnings can never be more than recommencements or occultation.[14]

In this chapter I follow the lead of Zygmunt Bauman.[15] My intent is not to offer any new account of Cambodian history; rather, it follows the pioneering work of David Chandler, Serge Thion, Ben Kiernan, Steve Heder, and Karl Jackson, among others. My purpose is not to add to the specialists' knowledge-base, or to identify heretofore unknown facts of Cambodian history. Rather, I propose to re-present Cambodian history in such a way as to articulate the geographical imaginations that undergirded the mass killings of the Khmer Rouge. In so doing I do hope to demonstrate the relevance of *geography* and *geographical knowledges* to those specialists working on Cambodia specifically, and genocide more broadly.

But still, we are left with the questions of when—and where—to begin. And, given such beginnings, how do we proceed with our re-presentation? As to the first set of questions, one common starting point is the 1863 treaty whereby Cambodia became a protectorate of France. Other possibilities include the political demonstrations of July 1942 that took place in Phnom Penh. As Foucault would recognize, each of these "points of departure" have their own contexts and precedents. For my present purpose, I go outside of Cambodia, and not with a Khmer revolutionary. Instead, I begin with the Vietnamese nationalist, Ho Chi Minh, and his founding in 1930 of the Vietnamese Communist Party (VCP) in a small house in a working class district in Kowloon, on the Chinese mainland across from the British colony of Hong Kong.[16]

Regarding the second question—of how to proceed—I draw on theoretical insights provided by those scholars studying social movements and revolutions.[17] To some readers, my situating the Khmer Rouge as a social movement—and, specifically, as a *resistance* group—may seem rather odd. To make such an argument carries a certain amount of risk, particularly within the study of resistance and social movements. There is a tendency among some academics, for example, to privilege

---

13 Here, parallels can be drawn with the attempts by Holocaust scholars to identify when Hitler's "Final Solution" to exterminate the Jews was reached. See, for example, Christopher R. Browning, *Nazi Policy, Jewish Workers, German Killers* (Cambridge: Cambridge University Press, 2000).

14 Foucault, *Archaeology of Knowledge*, 22.

15 Bauman, *Modernity and the Holocaust*, xii–xiii.

16 William J. Duiker, *The Communist Road to Power in Vietnam*, 2nd edition (Boulder, CO: Westview Press, 1996), 32.

17 Forrest D. Colburn, *The Vogue of Revolution in Poor Countries* (Princeton: Princeton University Press, 1994); Sidney Tarrow, *Power in Movement: Social Movements and Contentious Politics*, 2nd edition (Cambridge: Cambridge University Press, 1998); Misagh Parsa, *States, Ideologies, and Social Revolutions: A Comparative Analysis of Iran, Nicaragua, and the Philippines* (Cambridge: Cambridge University Press, 2000); Mark N. Katz (ed.), *Revolution: International Dimensions* (Washington, DC: CQ Press, 2001); Max Elbaum, *Revolution in the Air: Sixties Radicals Turn to Lenin, Mao, and Che* (London: Verso, 2004).

or even idolize social and/or "resistance" movements. As Tim Cresswell explains, the "characteristic move of progressive social and cultural theorists has been to identify with and make moral investment in subaltern or subordinated groups. It is their actions that have been described as resistance—as contestations of dominant spatialised norms."[18] Is it possible to conceive of the Khmer Rouge as a *subaltern* or *subordinated* group? Should (or even, could) the Khmer Rouge be viewed as a *resistance* group, when it's members subsequently detained, tortured, and killed innocent people, when it was the Khmer Rouge leaders who fashioned a genocidal program to promote their radical political agenda?

I am not asking readers to "identify" with, or make a "moral investment" in, the Khmer Rouge. Nor I am suggesting a sympathetic reading of their actions. However, I am arguing that just as it is imperative to understand how perpetrators viewed social groups as "lives not worth living," so too is it imperative to understand how the perpetrators viewed their own existence as a movement. And indeed, *from the perpetrators point of view*, they were a resistance group—revolutionaries— struggling against foreign imperialists and a corrupt monarchy. For individuals such as Pol Pot, Khieu Samphan, and Noun Chea, the Khmer Rouge was resisting Sihanouk, the French, the Americans, Lon Nol, Vietnamese communist dominance. And for many of the Khmer Rouge, they expressed a deep and profound belief in the righteousness, the justness of their actions. And in this manner they differed little from other perpetrators of mass violence. From the perspective of the Khmer Rouge leaders, they were part of an heroic social movement, a resistance against domestic and foreign injustices. Hence, I concur with Cresswell's argument that we should view resistance not as a potent symbol of subaltern freedom, but as an indicator and diagnostic of power.[19]

And now, to the emergence of the Khmer Rouge.

The Construction of Revolution in Indochina

Having just spent four years working with Vietnamese exiles in Siam, Ho Chi Minh had arrived in Kowloon in December of 1929. The occasion of his visit was to bring order to the chaos that presently engulfed the fledgling Communist movement in Vietnam. Owing to regional factionalism and personality clashes, the decade-long efforts of Ho Chi Minh were in jeopardy.

Born Nguyen Sinh Cung on May 19, 1890, in the coastal province of Nghe An, Ho Chi Minh would emerge as the foremost nationalist figure of Vietnam. The son of an educated, but relatively impoverished mandarin, Ho was tutored on Confucian philosophy and patriotic stories of Vietnamese nationalism. He briefly attended the Imperial Academy at Hue, but abandoned his lessons after one year. Later, following a short employment as a language instructor, Ho left Vietnam at the age of 19 and spent several years at sea. Working as a cook on a merchant ship, Ho traveled widely,

---

18 Tim Cresswell, "Falling Down: Resistance as Diagnostic," in *Entanglements of Power: Geographies of Domination/Resistance*, Joanne P. Sharp, Paul Routledge, Chris Philo, and Ronan Paddison (eds) (New York: Routledge, 2000), 256–68; at 258.

19 Cresswell, "Falling down," 258.

gaining first-hand experience into the lives of workers the world over. He visited port cities and other locales throughout Africa, Asia, and Europe. Later, during the First World War, Ho lived in London, and then Paris. In 1919, now known as Nguyen Ai Quoc (Nguyen the Patriot), Ho presented a list of demands to the leaders of the great powers at the Paris Peace Conference. In his appeal at the Treaty of Versaille, Ho submitted an eight-point program demanding Vietnamese elections for representation in the French parliament; freedom of speech, press, and association; release of political prisoners in Indochina; and equality of law for the Vietnamese with the French.[20] Although his demands were ignored by the European leaders, Ho was becoming widely known as a nationalist and opponent of colonialism.

Personally, Ho's objective was to end French colonial rule in Vietnam. How this was to be accomplished was secondary. It was only through his travels and his readings that Ho eventually turned to Communism. In 1920, for example, Ho read Lenin's "Thesis on the National and Colonial Questions," and from this vantage point, Ho began to interpret global affairs through the prism of Marxist-Leninist terms. In Paris he became a founding member of the Communist Party of France (CPF) and spoke often on the conditions of oppressed colonial peoples. It was through these efforts that he attracted the attention of the Comintern and was pegged to serve as catalyst for Communist activities in the French colonies of Indochina.[21]

Ho's activities are vitally important for understanding the Cambodian Communist movement, in that the liberation struggles of both Vietnam and Cambodia[22] are intimately associated with French colonial rule. And from this perspective, the emergence of revolutionary ideologies throughout Indochina was part of a broader— both in terms of history and geography—anti-colonial movement that swept through the former subjugated territories of the world.

The arrival of European missionaries and merchants throughout the seventeenth-century threatened the territorial integrity of Vietnam. Traders and merchants, for example, had begun to establish trading posts to take advantage of local business opportunities, as well as to participate more directly in trade with China. Catholic missionaries, largely from France but also Spain, also began to set up their activities in the region. By the end of the eighteenth-century, French missionaries claimed about 600,000 converts in southern Vietnam and about 200,000 in the northern regions.[23]

Vietnamese leaders, in particular, viewed the spread of Christianity as a threat to Vietnamese political authority. Large-scale persecution of converts and missionaries began in the 1820s and continued through the 1850s. French missionaries were routinely arrested and deported, while Vietnamese Catholic converts were jailed or, at times, executed. Catholic missionaries, in turn, demanded that France take

---

20 Robert D. Schulzinger, *A Time for War: The United States and Vietnam, 1941–1975* (New York: Oxford University Press, 1997), 9.

21 Duiker, *Communist Road to Power*, 16.

22 The Laotian liberation struggle is likewise connected, but has less direct relevance on events in Cambodia.

23 D.R. SarDesai, *Vietnam: Past and Present*, 4th edition (Cambridge, MA: Westview Press, 2005), 34.

military action. Only belatedly, when French merchants and manufacturers demanded protection of their own overseas commerce activities, was the intransigent French government moved to action.[24]

In 1858 a joint Franco-Spanish naval force bombed the Vietnamese port city of Danang, supposedly to obtain religious freedom for Catholics. More pressing, however, was a goal to force the Vietnamese emperor to accept a greater French dominance of economic activities. Although Spain ended its role in the hostilities following assurances from the Vietnamese government that persecution of Catholics would cease, the French continued the fighting for another three years. Ultimately, in 1862, France forced a treaty from the Vietnamese emperor. As part of the agreement, France received three provinces in southern Vietnam, as well as assurances that no other part of Vietnam would be transferred to any other power. Furthermore, the emperor agreed to pay an indemnity of four million piasters in ten annual installments and to open three ports in central Vietnam to French trade. France, lastly, was given permission to navigate the Mekong River.

For the neighboring kingdom of Cambodia, French colonialism would prove both a blessing and a curse. In the mid-1850s the Khmer king Ang Duang (r. 1848–60) sought French support, ironically, in an effort to retain Khmer independence. By the nineteenth-century, Cambodia was but a remnant of its former self, the glory years of Cambodia long since forgotten. Between the ninth and fourteenth centuries Khmer kings had reigned over vast territories that included much of present-day Cambodia, Thailand, Burma, Laos, and Vietnam. The reign of Suryavarman II (r. 1113–50), for example, stretched from central Vietnam in the east, to the Irrawaddy River in the west. From the fifteenth-century onwards, however, internal decay and foreign encroachment from neighboring Siam (present-day Thailand) and Vietnam combined to threaten Cambodia's very survival. Assorted Khmer kings made various attempts to secure the sovereignty of their kingdom. In 1593, for example, King Sattha (r. 1576–94) asked the Spanish governor of the Philippines for assistance in halting Siamese advances. This arrangement failed, however, and in 1594 Thai forces captured the Khmer capital (then located at Lovek, north of Phnom Penh). A new capital was located at Oudong and all subsequent Khmer monarchs were required to enter into vassal relationships with Siam. Similar relationships were demanded by their Vietnamese neighbors.[25]

Such was the historical precedent of King Duang's request. Threatened from his eastern and western neighbors, the Khmer king looked to the French for help. In 1853, encouraged by French missionaries, Duang sent a letter to Emperor Napoleon III asking for assistance. It is unknown if Napoleon ever responded. Three years later, however, a French diplomatic mission arrived to negotiate a commercial treaty with the Khmer king. This draft treaty failed to reach the Cambodian court.

Having made inroads into Vietnam, French officials were captivated by the possibilities Cambodia seemed to offer. In particular, the French saw in Cambodia

24 Patrick J. Hearden, *The Tragedy of Vietnam: Causes and Consequences*, 2nd edition (New York: Pearson Longman, 2005), 5–6.

25 Donald M. Seekins, "Historical Setting," in *Cambodia: A Country Study* (ed.) R.R. Ross (Washington, DC: US Government Printing Office, 1990), 3–71; at 13–14.

a means to access the much-desired markets of southern China and the possible sources of riches. Their interests were piqued by the well-publicized travels of French naturalist Henri Mouhot. It was Mouhout who, between 1859 and 1861, had journeyed up the Mekong River and "discovered" the ruins of Angkor outside of Siem Reap. His subsequent narratives of Angkor suggested that Cambodia contained vast treasures that were ripe for the taking.[26] Following the death of Duang in 1860 French officials worked to conclude a treaty with the king's successor and eldest son, Norodom (r. 1860–1904). As a result of these efforts, in 1863, in exchange for timber concessions and mineral exploration rights, Cambodia was declared a French protectorate.

The colonization of Cambodia was part-and-parcel of a larger French movement on the Asian mainland. Over the next two decades, between 1883 and 1885, French forces conquered the remaining territory of Vietnam and eventually declared protectorates over northern Vietnam (Tonkin) and central Vietnam (Annam). Concurrently, France claimed the territories of present-day Laos. By 1893 the Union Indochinoise (Indochina) had been formed, consisting of Tonkin, Annam, Cochin China, Laos, and Cambodia.

French colonial policy was quixotic. On the one hand, French officials forwarded a policy of "assimilation." In an attempt to bridge the gap between humanitarianism—as exemplified by the French concept of *mission civilisatrice*, or "civilizing mission"—and the harsh realities of colonialism, French adminstrators sought to remake Indochina in the image of France. French culture and society was transferred, in limited and selective ways, to the peoples of Vietnam, Cambodia, and Laos. Education, for example, when provided, was conducted in French, and the students learned *not* Vietnamese or Khmer history and geography, but rather French history and geography. On the other hand, French officials established a policy of "association" whereby French administrators would work with native leaders to concentrate on economic policies.[27]

Indochina was to provide France with a means of stabilizing its domestic economy. Most French officials favored a strong program of overseas economic expansion but such efforts needed to be tempered by the high initial costs of establishing colonies. Consequently, stemming from a concern among French officials that their colonial endeavors would result in a heavy tax burden on their own citizens, it was agreed that these costs would be transferred to the colonial subjects.[28] In other words, the costs of colonial administration would be supported by the very people who were colonized.

A new tax system was imposed in Indochina, one that ensured that colonial subjects would pay for their own subservience. First, three local budgets were formed, based on the direct taxation of residents of the three colonies. These budgets were then used to defray the costs of developing the entire colonial structure. Additional revenue was generated by placing a high tariff on goods imported into the colonies and by organizing state-controlled monopolies that sold licenses for the production

---

26 Seekins, "Historical Setting," 14–16; Chandler, *A History of Cambodia*, 140.

27 Tucker, *Vietnam*, 36–37.

28 Heardon, *Tragedy of Vietnam*, 8.

and distribution of opium, alcohol, and salt. Most of these funds were toward the construction of a colonial infrastructure (e.g., bridges, railroads, and harbors).[29]

The French colonial government also raised agricultural taxes and imposed a head tax that had to be paid in cash. Consequently, this policy forced many peasant families to send their husbands, fathers, and sons to the cities, mines, and plantations in search of waged labor. Through this process the French created their own supply of surplus labor, one that was exploited for work in coal mines, rubber plantations, and the construction of railroads.[30]

The French emphasis on making Indochina self sufficient, and self supporting, contributed to pervasive—albeit uneven—changes in local societies. On the one hand, Indochina as a whole would provide a source of raw materials. Manufacturing, though, was largely neglected. Rubber was promoted, and large quantities of raw latex were exported to French where it was converted into tires and other finished products. In turn, these products were exported *back* to Indochina, generating profits for French domestic industries. Furthermore, France sought to prevent foreign competitors from exporting other manufactured goods into Vietnam, Cambodia, or Laos.[31] On the other hand, French authorities approached the regional components of Indochina very differently. French penetration in Tonkin, for example, was rather minimal. In the south, however, Vietnamese society and economy were dramatically altered. From the outset, French colonial administrators in Cochin China adopted a narrow fiscal objective of balancing the colonial budget, paying the ever-expanding French administrative personnel, and, if possible, creating a profitable economy that would justify the costs of colonization. A major vehicle was the cultivation of rice for export.[32]

Previously, the export of rice was forbidden by Vietnamese emperors, in large part to protect against times of famine. Furthermore, prior to colonization, all lands were owned by the emperor. Decisions were made through village councils, rather than individual land-owners. French colonial officials, however, reconfigured farming practices in the south through the introduction of private ownership and the establishment of large land-holdings. In Cochin China, unlike Tonkin, land was plentiful and subsequently was given in the form of concessions to French nationals or "deserving" Vietnamese citizens who had loyally served the French. Alternatively, lands were auctioned off in large lots to off-set the cost of canal building in the Mekong Delta. Over time, Vietnamese landlords—a new artistocracy—became extensions of the French colonial government.[33] In Cochin China, a mere 6,316 owners controlled 45 percent of the total cultivated lands. In effect, French colonialism transformed

---

29 Hearden, *Tragedy of Vietnam*, 8.

30 Jeffery M. Paige, *Agrarian Revolutions. Social Movements and Export Agriculture in the Underdeveloped World* (New York: The Free Press, 1975), 283.

31 Peter Church, *A Short History of South-East Asia*, revised edition (Singapore: John Wiley & Sons, 2003), 190; Hearden, *Tragedy of Vietnam*, 9.

32 Paige, *Agrarian Revolutions*, 281–283

33 Paige, *Agrarian Revolutions*, 281.

the old class of feudal landlords and imperial officials into a new class of capitalist landlords and colonial officials.[34]

In Cambodia, by contrast, French colonial practices were less disruptive. To be sure, French authorities assumed full control over Cambodian affairs, including financial, legal, commercial. French officials also heavily taxed the Cambodian people and large percentages of rice harvests were confiscated, contributing to high levels of rural indebtedness. A bureaucratic hierarchy was established, with French nationals occupying the higher echelons of government, with ethnic Vietnamese placed in the lower rungs of government. Cambodians were mostly excluded. The Cambodian monarchy was reduced to that of a figure-head. The French also established a modest transportation system, but one that benefitted the exploitation of natural resources. They otherwise did very little to modernize or improve the Cambodian economy. Large plantations were established for the cultivation of rubber and other agricultural products.[35]

The Genesis of the Indochinese Communist Party

In Vietnam, as in other agrarian societies, peasant rebellions against starvation and government misrule were not infrequent. Even in the best of times, daily life was precarious, seemingly at the whim of rain or drought, floods or pestilence, natural conditions that often meant the difference between food or famine. In colonial Vietnam, however, these conditions were worsened by high taxes, monopolies, seizures of communal lands, and forced labor policies.[36]

According to William Duiker, the many peasant rebellions that erupted throughout Vietnam's countryside in the late nineteenth and early twentieth-centuries were primarily social and economic in nature. And although an element of anti-French sentiment may have been present, these uprisings cannot realistically be considered to be anti-colonial in intent. Indeed, for the most part, the rural villages of Vietnam were politically quiescent.[37]

The cities of Vietnam told a different story. Particularly in Cochin China, an increasing number of French-educated Saigon intellectuals began demanding greater opportunities. According to Mark Bradley, the "sons, and sometimes the daughters, of mandarin families who formed the new intelligentsia of the 1920s were products of the French-controlled educational system—and especially its veneration of Western ideas and values—that superseded the traditional academies that had taught the Chinese classics and Confucian morality, generated profound changes in the attitudes and ideals of these Vietnamese." Whether seeking reform or revolution, members of the commercial bourgeoisie in general sought equal economic and

34 Kevin Cox, *Political Geography: Territory, State, and Society* (Malden, MA: Blackwell, 2002), 304; Jonathan Neale, *A People's History of the Vietnam War* (New York: The New Press, 2003), 13; Tyner, *America's Strategy in Southeast Asia*, 75–77.

35 Evan Gottesman, *Cambodia After the Khmer Rouge: Inside the Politics of Nation Building* (New Haven, CT: Yale University Press, 2003), 15.

36 Duiker, *Communist Road to Power*, 10.

37 Duiker, *Communist Road to Power*, 10.

political rights with their French and Overseas Chinese competitors. Subsequently, by the 1920s a number of nationalist organizations began to appear in all three divisions of Vietnam. In Tonkin, an assortment of teachers and civil servants formed the Viet Nam Quoc Dan Dang (VNQDD, or Nationalist Party). Modeled after Sun Yat-sen's Kuomintang (KMT, or Nationalist Party), the VNQDD was extremely anti-French in its outlook and, strategically, viewed violent revolution as the most promising course of action. In Annam a milder anti-French organization emerged. Known as the Tan Viet (New Revolutionary Party), this organization was divided between the pursuit of violent or non violent methods. And lastly, in Cochin China, a region heavily transformed by French policy and thus most ripe for revolution, there emerged many competing parties and organizations, most notably the "Hope of Youth" (Thanh Nien Cao Vong) Party, founded by Nguyen An Ninh.[38]

As these organizations were being established, Ho Chi Minh returned to Vietnam to form his own revolutionary organization, one that would be founded on Marxist-Leninist principles. Having attracted the attention of the Comintern, in 1923 Ho Chi Minh attended the University of the Toilers of the East in Moscow and in 1924 he was a leading spokesperson for the anti-colonial cause at the Fifth Congress of the Communist International.[39] Recognizing the political "innocence" of the Vietnamese people—and especially the peasants—Ho Chi Minh promoted a goal of national liberation rather than communist revolution. This was both an ideological and pragmatic reason. On the one hand, Ho Chi Minh was a dedicated nationalist, who desired an independent Vietnam free of foreign domination. Communism, for Ho Chi Minh, provided the most viable path. On the other hand, Ho Chi Minh also understood that many Vietnamese might not support the socialist ideals promoted by a Communist revolution. Ho Chi Minh required a united front, one that included those on the political Left as well as the Right, to engage in the anti-colonial struggle.

In 1925 Ho Chi Minh formed the Vietnamese Revolutionary Youth League (Viet Nam Thanh Nien Cach Mang Dong Chi Hoi). He contacted various Vietnamese groups and recruited members from a wide array of progressive classes. And though he personally sought to erect in Vietnam a society based on Marxist principles, he acknowledged that this was a long-term goal. The immediate task was also the liberation of Vietnam from colonial dominance.

Drawing on his experiences at the University of the Toilers, Ho Chi Minh encouraged members of the Vietnamese Revolutionary Youth League to travel to China, where they learned about capitalism and Marxism. Although some of these students remained and worked in China, or pursued further studies in the Soviet Union, most returned to Vietnam and participated in subsequent recruitment campaigns to

---

38 Duiker, *Communist Road to Power*, 11–12.

39 Mark P. Bradley, *Imagining Vietnam & America: The Making of Postcolonial Vietnam, 1919–1950* (Chapel Hill: The University of North Carolina Press, 2000), 33. The University was founded on Lenin's order in 1921 and was originally placed under the jurisdiction of Joseph Stalin's People's Commissariat of Nationalities. Under Stalin's direction, the school became the leading institute for training Asian revolutionaries invited to Soviet Russia. There were over 150 instructors, teaching courses on the natural and social sciences, mathematics, revolutionary history, and Marx's theory of historical materialism. See William J. Duiker, *Ho Chi Minh: A Life* (New York: Theia, 2000), 92–94.

build the Revolutionary Youth League. Within two years Ho's organization became an active political force and a formidable rival for other organizations in Vietnam.[40]

Ho Chi Minh well understood the fundamental problem of building a Communist movement. A sizeable number of potential members were driven by anti-French sentiments; political, social, and economic liberation were the goals, and not necessarily a Communist society. Consequently, many members were driven by nationalist rather than socialist ideologies. Such ideological differences threatened to fracture any organization or movement, including the Revolutionary Youth League. And indeed, while Ho Chi Minh was attempting to organize Communist activities in Siam, the League was rapidly disintegrating through internal dissension.

Soviet leadership decided that nationalism and Communism were incompatible. Having witnessed the violent sundering of nationalist/Communist alliances in both China and the Dutch East Indies, Soviet officials mandated that all Party activities were to be strictly Communist. And in Vietnam, the radical Tran Van Cung set out to transform the Revolutionary Youth League into a full-fledged Communist Party. At a meeting held in Canton in 1929, Tran, along with a number of delegates from Tonkin, made a formal proposal to disband the League and to establish a Communist Party. Other delegates, indeed the majority, rejected this proposal. Tran left in protest and on his return to Hanoi he formed the Indochinese Communist Party (ICP, or Dong duong Cong san Dang). Over the next few months, as more radical members joined the ICP, leaders of the Revolutionary Youth League realized their tactical error. Accordingly the League was transformed and renamed into a competing Communist Party, known as the Annam Communist Party (ACP, or Annam Cong san Dang). Concurrently, radical members of the Tan Viet Party renamed their organization the Indochinese Communist League (Dong duong Con san Lien doan). Remarkably, Vietnam now was home to three competing Communist parties.[41]

Neither Comintern officials nor Ho Chi Minh were pleased with the events transpiring in Vietnam. Thus, to prevent the fledgling Communist movement from collapsing entirely, Ho Chi Minh met with representatives from the ICP and the ACP. Meeting in Kowloon, Ho stressed the necessity of unity, and urged the members to overcome their regional differences. He further suggested that both parties be dissolved, to be replaced by a new formation called the Vietnamese Communist Party (VCP). To this, the delegates readily agreed. Ho Chi Minh further proposed that the VCP accept any Vietnamese who supported the overthrow of the French: nationalists, petty bourgeoisie, even rich peasants and small landlords.[42]

Months later, at a meeting not attended by Ho Chi Minh, the VCP was transformed yet again. This time, under instructions from Moscow, the party was renamed the Indochinese Communist Party (ICP, or Dang Cong san Dong Duong). Not to be confused with Tran's former party, this new ICP was envisioned to enable the various territories of French Indochina to be approached as a single unit. At the urging of the Comintern, it was decided that the three peoples—the Lao, Khmer, and the Vietnamese—of Indochina should be united both during and after the anti-

40 Duiker, *Communist Road to Power*, 18.
41 Duiker, *Communist Road to Power*, 30–31.
42 Duiker, *Communist Road to Power*, 32–33.

French struggle. From the perspective of Comintern, French colonial rule had already imposed a framework of unity among Indochina and that this should serve the revolution.[43]

The key objective, from the perspective of the Comintern, was to move the entire revolutionary movement away from a singular focus of anti-colonialism, to one of anti-feudalism.[44] Ho Chi Minh strongly disapproved of the name change and the underlying principle. On the one hand, French rule had been uneven throughout Indochina. Consequently, conditions for revolution were vastly different throughout the entire region. Not only were conditions different between Cambodia and Vietnam, so too were conditions different between Tonkin, Annam, and Cochin China. In short, Vietnamese revolutionaries such as Ho Chi Minh understood and appreciated the "geography" of the situation much better than their Comintern mentors. On the other hand, the Comintern perspective also failed—unlike Ho Chi Minh—to consider the longer histories of political activity in Indochina. And in fact these regional differences, and especially the historical animosity between Vietnamese and Khmer, would continue to plague the broader anti-colonial movement for decades to come. Nevertheless, the ICP remained in name and spirit and would become the vehicle for Communist activity throughout all of Indochina, including Cambodia.

Vietnamese leaders were reluctant to assume the role mandated to them by the Comintern, largely because they were doubtful of the overall success of any revolution in Cambodia. The Vietnamese leaders did, however, see in Cambodia an important site in their own anti-colonial and Communist revolution. Furthermore, most Vietnamese Communists readily agreed that all resistance movements in Indochina should contribute to the Vietnamese revolution. In fact, Vietnamese party leaders interpreted the principle of Indochinese unity as strategic in character. The elimination of French rule in Vietnam *must* come with the elimination of French rule in *both* Laos and Cambodia.

Vietnam and the Cambodian Stage

Revolutionary movements must be manufactured. The "ordinary" peasant or proletariat, despite Marx's initial thought, does not simply rise up to overthrow the existing order. Indeed, the relationship between "colonialism" and "liberation" is complex. For the majority of colonial subjects, it may not always be self-evident that an alternative to the present colonial structure exists, nor how an alternative if conceived could be pursued. Colonial subjects may also—and with good reason—fear reprisals from colonial authorities.[45]

Cambodia at the dawn of the twentieth-century exhibited certain pre-conditions that provided some impetus for revolution. French colonial control was repressive and discriminatory. However—and this is an observation made at the time by

43 Gareth Porter, "Vietnamese Communist Policy Toward Kampuchea, 1930–1970," in *Revolution and its Aftermath: Eight Essays*, David P. Chandler and Ben Kiernan (eds) (New Haven, CT: Yale University Southeast Asia Studies, 1983), 57–98; at 58.

44 Duiker, *Communist Road to Power*, 40.

45 Colburn, *Vogue of Revolution*, 42.

Vietnamese members of the ICP—anti-French sentiment in Cambodia was not as pervasive as it was in other French possessions. In fact, Cambodia was relatively quiescent politically during the first four decades of the twentieth-century. Certainly, some rebellions occurred throughout the late nineteenth and early twentieth centuries, but none of these could be considered part of a large, anti colonial movement.

Numerous explanations, both then and now, have been forwarded to account for the paucity of anti-French sentiment and action in Cambodia. In part, the existing (and small) cohort of Khmer educated elites in Cambodia had little political experience. French authorities had given most (and certainly the most important) administrative jobs to the Vietnamese; authorities had also encouraged Vietnamese immigration to the cities (where the new arrivals assumed most of the economic functions); and brought Vietnamese laborers into the rubber plantations. Consequently, most early resistance movements in Cambodia were led by, and composed of, Vietnamese workers.[46] In fact, during the 1930s a cell of Ho Chi Minh's Indochina Communist Party (ICP) was established. Operating mostly in the rubber plantations, these cells were composed primarily of ethnic Vietnamese laborers, with only a few Khmer.

The carefully maintained fiction of royal rule was also a major factor. Although in practice the king was reduced to little more than a figure-head, symbolically the peasants could see the king occupying the Khmer throne. Relatedly, David Chandler suggests that radical social action, like ideas of a just society, was less prestigious and widespread in Cambodia than in Vietnam because Khmer society viewed political affairs not as the people's business, but as royal business.[47] Low literacy rates—which the French were reluctant to improve—also insulated the majority of the population from the nationalist and anti-imperialist ideas that were circulating throughout other parts of Indochina.[48]

Required for the emergence and continuance of a social movement, let alone a revolutionary movement, is the existence of a "vanguard" that may effectively align disparate elements of society. As Forrest Colburn explains, while the origins of contemporary revolutions are clearly rooted in social, political, and economic conflict, the outcomes of these revolutions have been determined by the political imagination of revolutionary elites.[49] In Cambodia, as we've already seen, such activities were initially propelled from Vietnam and Siam (renamed Thailand in 1939). In fact, it is somewhat ironic that just as Siam and Vietnam threatened to obviate the existence of Cambodia, so too did these political neighbors provide early inspiration and guidance in Cambodia's anti-colonial movement. Such influence, however, did not come without a price. As discussed later, continued mis-trust between Khmer revolutionaries and their neighbors would influence just how far this external guidance would go. For many Khmer revolutionaries, it made no sense to replace one colonial power (France) with another (Vietnam or Thailand).

---

46 Porter, "Vietnamese Communist Policy," 60.

47 Chandler, *Tragedy of Cambodian History*, 4.

48 According to Ian Brown, the "French justified the lack of investment in education by claiming dishonestly that Cambodians themselves did not want schools and colleges." See Ian Brown, *Cambodia* (Oxford: Oxfam, 2000), 22.

49 Colburn, *Vogue of Revolution*, 5.

In a further irony, French scholarly activities contributed to the rise of anti-colonial and nationalist sentiments among the Khmer. Fascinated with the monumental structures found in Siem Reap province, namely Angkor Wat and Angkor Thom, French archaeologists wrote profusely on the glories of the Angkorean empire, as well as its decline. Following Norodom's ill-fated agreement with, and concession to, France in 1862, the French effectively constructed Cambodia's people and past. As Chandler explains, the French were struck by the contrasts they perceived between the grandeur of the Angkorean ruins on the one hand and, on the other, the "decline" they saw in nineteenth century Cambodia. He contends that the French invented Cambodia as a charming, powerless protectorate and subsequently took control of Cambodia's historical narrative. This French-constructed knowledge, which Chandler terms "Cambodge," "blended the grandeur of Cambodia's past, symbolized by the Angkor ruins, with an assessment of the Cambodian people as insouciant and needful of protection."[50]

An unanticipated consequence of these writings, however, was that the historical narrative constructed by French scholars generated a "crisis of identity" among many of the more-educated Khmer in the 1930s and 1940s. In particular, Cambodian students were confronted with the contradictions of past and present. They could not but compare the impotence of current kings with the strength of the former kings of Angkor. In this way, French writings served as a catalyst for Cambodian nationalism, including that of the communist movement.[51] As Chandler explains, "France constructed a seductive, unserviceable heritage for their proteges, who were told that they were simultaneously needful of protection and the worthy descendants of the kings of Angkor." However, Chandler suggests that while such narratives may at times keep the population docile, these lessons would occasionally promote, especially among political leaders after independence, a severe case of *folie de grandeur*.[52] Throughout his reign, for example, Sihanouk repeatedly would compare himself to the Angkorean monarchs, as would his successor, Lon Nol. And even Pol Pot proudly declared that "If our people can build Angkor, they can do anything." These men were, in the words of Chandler, "entranced by the 'otherness' of their own past, eager to compensate for colonially induced powerlessness, and entranced by what they perceived as Cambodia's incomparability."[53]

Associated with the impact of French scholarship was a key institutional development. In 1930 the Buddhist Institute was established by the French, and administered by a French scholar-administrator named Suzanne Karpelès. Located in Phnom Penh, and only a few hundred yards from the royal palace, the Institute "acted as a clearing house and repository for Cambodian religious and literary texts and as a meeting place for Buddhist intellectuals." More significantly, the institute also exposed the small, but growing class of Cambodia's intellectuals—most of whom were either monks or former monks—to some of the political currents

50 Chandler, "From 'Cambodge' to 'Kampuchea': State and Revolution in Cambodia, 1863–1979," *Thesis Eleven* 50(1997): 35–49; at 35.

51 Chandler, *Brother Number One*, 13.

52 Chandler, "From 'Cambodge' to 'Kampuchea,'" 37.

53 Chandler, "From 'Cambodge' to 'Kampuchea,'" 38.

affecting the Khmer minority in Cochin China.[54] Two of these monks were Son Ngoc Thanh and Achar Sok.

Born in 1908 to a Khmer father and a Sino-Vietnamese mother in southern Vietnam, Son Ngoc Thanh lived in relative prosperity. His earliest education was in Khmer-language pagoda schools both in his home-town of Travinh and in Saigon. Son Ngoc Thanh subsequently received a secondary education in France, completing a teaching diploma while also studying law at the University of Paris. In 1933, back in Cambodia, Thanh briefly worked at the Phnom Penh Library before becoming a magistrate in Pursat and then a public prosecutor in Phnom Penh. It was Karpelès who brought Son Ngoc Thanh to the Buddhist Institute, where he became the Institute Secretary. In 1936, while at the Institute, Son Ngoc Than, along with Pach Chhoeun, established the country's first Khmer-language newspaper, *Nagara Vatta* (Pali, for Angkor Wat). Although the newspaper was not explicitly anti-French, it did print occasional stories critiquing French colonial rule. More important was that Thanh, through both the Institute and the newspaper, was able to raise national awareness, thereby stoking the fires for independence.[55] Some of the more specific demands made by Thanh included greater participation of Cambodians within commerce, greater educational opportunities, and equal treatment under the law for Cambodian citizens.

Working alongside Son Ngoc Thanh at the Buddhist Institute was another monk named Achar Sok. Raised in southern Vietnam and fluent in Vietnamese, Achar Sok would later join the ICP in 1945 and, through the 1950s become Saloth Sar's patron. He would later assume the name Tou Samouth.[56] During the late 1930s and early 1940s Achar Sok, along with Son Ngoc Thanh, Pach Chhoeun, and various other monks, began visiting wats throughout the country, calling for an intellectual re-awakening of the Khmer.[57] These efforts, however, were forestalled with the advent of war.

The Japanese Interlude

The Second World War provided a catalyst for Cambodian nationalism. In the summer of 1940 France was militarily defeated by Nazi Germany. Subsequent agreements brought about an armistice that promised French sovereignty despite that country's being occupied by Germany. A new government was proclaimed, named after the site of its administrative center in Vichy. As part of the armistice, Vichy France would continue to control the French Empire, including its colonies in Indochina. French officials hoped that they could simply endure the war and, no matter the outcome, emerge with most of their territories (both in France and in their colonies) intact. Consequently, in August 1940, following the lead of the collaborationist Vichy government in France, the governor-general of Indochina signed a general accord

---

54 Chandler, *Brother Number One*, 16.

55 Elizabeth Becker, *When the War was Over: Cambodia and the Khmer Rouge Revolution* (New York: Public Affairs, 1998), 42.

56 Chandler, *Brother Number One*, 24.

57 Kiernan, *How Pol Pot Came to Power*, 23.

with Japan, which allowed the French administration to continue in Indochina in return for placing the military facilities and economic resources at Japan's disposal.[58] By September 1940 Japanese troops moved into northern Indochina and, by May of the following year, occupied much of Cambodia.

Throughout the ensuing months of the war France's territorial integrity, particularly among its colonies, was greatly reduced. In late 1940, for example, the pro-Japanese government of Phibunsongkhram[59] (Phibun), the Thai Prime Minister, aware of French military weaknesses, sought to regain territories in Cambodia and Laos that the Thai had ceded earlier in the century to the French.[60] Consequently in November the Thai army invaded northwestern Cambodia. In turn, the Japanese government, overseer of France's Southeast Asian colonial possessions intervened, supported Thailand and forced the French to cede the province of Battambang and most of Siem Reap province. Angkor was to remain under French control. The loss of Cambodian territory humiliated King Monivong and the king retired to his farm in Kompong Chhnang. During the last remaining months of his life he refused to meet with French officials or even to speak the French language.[61]

Unlike other colonies, such as the Philippines, Cambodia was not physically devastated by events of the Second World War. The political environment of Cambodia was a different story. Although overshadowed by the Japanese, French officials continued to dictate Cambodian developments, including political successions. When King Monivong died the French chose Norodom Sihanouk—the great-grandson of King Norodom—as the next monarch. Although Prince Monireth, King Monivong's son, was considered the logical heir apparent, French authorities believed instead that the younger and inexperienced Sihanouk would best serve their interests. And indeed, throughout most of the Second World War, Sihanouk spent his time in Phnom Penh and offered no visible resistance to either the Vichy French or Japanese government.[62]

---

58 SarDesai, *Vietnam*, 56.

59 Born Plaek Khitasangkha (or Khittasangkha) near Bangkok in 1897, Phibun's upbringing was militarily-based. At the age of twelve he attended military academy and, upon graduation, entered the artillery. Based on his service record, he earned advanced military training in France, during which time he became a leader of a student movement that was plotting a military overthrow of the monarchy. He returned to Siam and was granted the rank and title, Luang Phibunsongkhram (which he later assumed as his family name). In 1934 he was made minister of defense, before assuming the post of prime minister in 1938—a position he held until 1944, and again from 1948–1957. It was Phibun who, in 1939, pushed through the change of the name of Siam to Thailand. See David K. Wyatt, *Thailand: A Short History* (New Haven, CT: Yale University Press, 1984), 252–254; see also Chris Baker and Pasuk Phongpaichit, *A History of Thailand* (New York: Cambridge University Press, 2005), 124–125.

60 Chandler, *History of Cambodia*, 166.

61 Chandler, *Brother Number One*, 17. King Monivong became seriously ill and died in April 1941.

62 William Shawcross, *Sideshow: Kissinger, Nixon, and the Destruction of Cambodia*, revised edition (New York: Cooper Square Press, 2002), 47.

Other Cambodians were not so pliable. Son Ngoc Thanh, who for years had opposed French rule, closely monitored the events of the early 1940s. Emboldened by the defeat of French forces in both Europe and Indochina—including the humiliating losses incurred by French forces to the Thai military—Thanh and his colleagues actively pursued a pro-Japanese and anti-colonial strategy. The stage was set for one of Cambodia's most defining events throughout the revolution.

On July 18, 1942 the French arrested two monks, Achar Hem Chieu and Achar Nuon Duong. Both men were accused of promoting anti-French sentiment and of plotting a coup d'état. Prior to arresting the monks, however, French authorities failed to defrock the monks and remove them from the *sangha*. These "inactions" were perceived by the Cambodian populace as both politically offensive as well as sacrilegious. In response, Son Ngoc Thanh and Pach Chhoeun organized an anti-French demonstration in support of the arrested monks.[63]

Two days after the arrests, Pach Chhoeun led a protest march of over one thousand people, with thousands more lining the streets. The participants marched through Phnom Penh to the office of the French *résident supérieur*. As Chhoeun presented a petition to French officials, he too was arrested. Within days he was tried, found guilty, and sentenced to death (which was later commuted to life imprisonment). Son Ngoc Thanh war more fortunate. He escaped arrest and went into hiding for several days in Phnom Penh, later making his way to the Thai-controlled city of Battambang. After a brief stop in Bangkok, Thanh eventually sought and received asylum in Japan, from where he struggled to keep the nationalist movement alive.[64]

Son Ngoc Thanh was not the only Cambodian to flee in fear of retribution. Another individual was Achar Mean. Born in Travinh around 1920 of a Cambodian father and a Vietnamese mother, Mean studied at Wat Unnalom in Phnom Penh. Following the monks' demonstrations in 1942 he left the capital where, at some point, he made contact with an ICP cell in Thai-held territory. In 1946, in an apparent attempt to capitalize on the reputation of Son Ngoc Thanh and to link himself to Ho Chi Minh, Achar Mean adopted the name Son Ngoc Minh.[65]

Achar Mean's initial contact with the Vietnamese was not happenstance. During the Second World War the Vietnamese Communists actively resisted the Vichy government. In May 1941 a new organization was established, the League for the Independence of Vietnam (Viet Nam Doc lap Dong minh, or Viet Minh). Downplaying communism, the Viet Minh was presented as a nationalist organization and all patriotic elements in Vietnam, regardless of class, were welcomed into the front against both the French and the Japanese. As part of their overall objectives, the Central Committee of the Viet Minh resolved that Japanese Facism and French imperialism were both inimical to the peoples of Indochina.[66]

Apart from the limited Vietnamese activities in Cambodia, other Khmer nationalists received sponsorship from the Thai government. In 1944, as the defeat of

---

63 Chandler, *Tragedy of Cambodian History*, 19.

64 Chandler, *History of Cambodia*, 168.

65 Chandler, *Tragedy of Cambodian History*, 34. For many scholars, Son Ngoc Minh is identified as the father of Cambodian communism.

66 Kiernan, *How Pol Pot Came to Power*, 40.

the Axis powers was becoming all too apparent, Phibun was forced to resign as prime minister. He was replaced by the left-leaning civilian, Pridi Phanomyong. Within the context of these political changes, in 1944 a Khmer resident of Thailand, Poc Khun, founded a movement called the Independent Khmer, or Khmer Issarak, which received support from Pridi's government. The purpose of the Khmer Issarak was to conduct anti-French propaganda in the Thai-controlled provinces of Battambang and Siem Reap in an attempt to foster the removal of the French from Indochina.

Overall, however, anti-French and anti-Japanese activities in Cambodia were minimal. For the "ordinary" Khmer peasant, life continued much as before. This is not to say that conditions were good. The Vichy regime, perhaps because of its recognized weaknesses, was "viciously repressive."[67] As Tully explains, the French police "rounded up thousands of real and imagined opponents and interned them in prisons and concentration camps." This "was also a period of mounting austerity with widespread shortages of food and clothing."[68]

For many of the more prosperous Khmer youth, the war-time experiences jostled uneasily with their on-going education. Despite the war, many Khmer students, particularly those from more well-to-do families and/or those with royal connections, were able to obtain higher education. Many students attended the *Collège Norodom Sihanouk* (a junior high school) which had recently been established and named after the young king. In 1942 twenty boys were selected as the first class, a cohort which included Saloth Sar. Others who attended would include Hu Nim, Khieu Samphan, and Hou Youn, all of whom would become important communist leaders. Classes were conducted in French and the boys studied literature, history, geography, mathematics, science, and philosophy. Significantly, the curriculum sowed the seeds of future nationalist sentiment. The students, for example, learned of France's "civilizing mission," as well as the glories of the French empire. However, the students also knew of, and indeed experienced first hand the decline of French hegemony. In Europe, France itself was occupied—in many respects, colonized—by Germany. And in Indochina, France suffered humiliating losses to Japan and Thailand.[69] This was most decidedly *not* the France that was taught in school.

A further lesson occurred in the waning months of the war. In the summer of 1944 Paris was liberated by the Allied powers, ending the Vichy government. In turn, Japan dissolved the French government in Cambodia. On March 9, 1945 Japanese forces arrested the French military, police, and native guards and eventually imprisoned the entire French civilian population.[70] Equally significant, the Japanese government encouraged King Sihanouk to declare independence.

With the French removed from power, Sihanouk hastily formed a seven-person cabinet. Significantly, this assemblage was decidedly *not* nationalist in orientation. None of the members, for example, had ever opposed the French and, indeed, most had occupied high positions in the colonial regime.[71] Nevertheless, between

---

67 Tully, *Short History of Cambodia*, 106.
68 Tully, *Short History of Cambodia*, 106–107.
69 Chandler, *Brother Number One*, 18–19.
70 Becker, *When the War was Over*, 48
71 Chandler, *Tragedy of Cambodian History*, 15.

March and December 1945, Cambodian authorities promulgated 155 laws and 390 administrative decrees. When the French returned to power, approximately three-quarters of the laws and nearly 90 percent of the decrees were allowed to stand. The most substantive laws and decrees, however, were overturned, notably those that proclaimed Cambodia's independence. Also overturned were two laws that decreed that income and property taxes should be collected—for the first time ever—from Europeans.[72]

Within this context, Son Ngoc Thanh, the veteran Khmer nationalist, returned to Cambodia after years in exile in Japan. Although absent from Cambodia since 1942, Thanh's reputation as a nationalist hero had grown considerably. In acknowledgment of this notoriety, in May of 1945 Thanh was named Cambodia's Foreign Minister.

The Reimposition of French Rule

During the summer months of 1945, it became clear to many Cambodians that Japan would ultimately lose the war. This had serious ramifications with respect to Cambodia's independence, for it was also clear that France would attempt to reimpose its colonial authority over the region.

On August 6, 1945 the United States dropped an atomic bomb on the Japanese city of Hiroshima and three days later, a second atomic bomb was dropped on the Japanese city of Nagasaki. On August 10, while Japanese authorities were still trying to comprehend the massive devastation unleashed by the atomic bombs, Emperor Hirohito counseled his cabinet to surrender. Japan formally surrendered on September 2.

In Cambodia, hard-line nationalists worried over the implications of Japan's defeat. Consequently, on August 9 a group of seven young Khmer officials stormed Sihanouk's palace in an attempt to force the prince to abdicate. Moreover, the rebels demanded the removal of "traditionalists" and their replacement with nationalists. According to Becker, these men wanted to put in place an established nationalist at the head of Cambodia, not a young collaborationist king who had proved little more than his ability to work with either French or Japanese overlords.[73] During the attempted putsch, the men arrested virtually the entire Cabinet, except for Thanh, who became Prime Minister the next morning.[74]

Ben Kiernan identifies three significant aspects of the coup. First, the move—supported as it was by many Buddhist monks and students—served as a clarion call to anti-colonial resistance. Second, the coup exposed internal rifts in Khmer nationalism. As a case in point, immediately after the coup Son Ngoc Thanh released the former ministers and arrested the seven coup leaders. The motivation of Thanh apparently hinges on his other efforts to solidify Cambodia's relations with its

---

72 Chandler, *Tragedy of Cambodian History*, 17.

73 Becker, *When the War was Over*, 49.

74 Kiernan, *How Pol Pot Came to Power*, 49.

neighbors, namely Thailand and Vietnam. Third, and related, Thanh recognized Ho Chi Minh's Democratic Republic of Vietnam and hoped to obtain its support.[75]

Worried most about the return of the French, Son Ngoc Thanh attempted to solidify Cambodia's position on the international stage. He hoped that through alliances with the Republic of China, Thailand, and Vietnam, Cambodia would be able to retain its sovereignty. Significantly, Thanh also attempted to strengthen relations between Cambodia and Vietnam. Many of these attempts met with little success. In September 1945, for example, Thanh dispatched a retired provincial governor to Bangkok to seek support for his regime. He hoped to gain cooperation from the newly installed anti-colonial government of Pridi Phanomyoung. However, given the tumultuous nature of Thai politics, Pridi himself was uncertain as to his own political fortunes and was thus unwilling to align himself with Son Ngoc Thanh.[76]

Son Ngoc Thanh also looked to the east. Thanh, unlike other Khmer nationalists, was not particularly anti-Vietnamese. But then again, neither was he a Communist. Nevertheless, he viewed in Vietnam a potential ally in the seemingly inevitable anti-French crusade. In September Son Ngoc Thanh was in contact with leaders of the Viet Minh who suggested that the two countries coordinate their resistance against the returning French. Son Ngoc Thanh accepted the proposal and dispatched a delegation to Vietnam. However, the Khmers made a strategic error in that they demanded, as a precondition for cooperation, the return to Cambodia of two Mekong Delta provinces, Soctrang and Travinh. This region had historically been settled by ethnic Khmers—in Cambodia, it was known as Kampuchea Krom, or "lower Cambodia." Vietnam, however, had authority over the two provinces and was unwilling to relinquish control. [77] Despite this initial failure, political developments in Vietnam would continue to reverberate throughout Cambodia.

On September 2, 1945 Ho Chi Minh had declared the formation of an independent Democratic Republic of Vietnam (DRV). However, the new republic was not immediately recognized by any country.[78] Through late 1945 and into 1946 French forces re-occupied Vietnam.[79] On March 6, 1946 a Preliminary Convention was signed in Hanoi. The French promised to recognize the government of the DRV as a free state within the French Union. Vietnam would have its own parliament, army, and finances, and would be part of an Indochinese Federation that included Cambodia and Laos. A referendum, furthermore, was scheduled to take place throughout Tonkin, Annam, and Cochin China to determine the final political status of Vietnam. This arrangement, however, preserved a French presence and left unclear the question

---

75 Kiernan, *How Pol Pot Came to Power*, 50. According to Kiernan, many of the coup leaders were able to escape; five would become active in the Indochina Communist Party. One alleged participant of the coup was Son Ngoc Minh. The betrayal of Son Ngoc Thanh partially explains why future cooperation between the two nationalists was difficult.

76 Chandler, *Tragedy of Cambodian History*, 24.

77 Kiernan, "Origins of Khmer Communism," 164.

78 SarDesai, *Vietnam*, 59.

79 Initially, Vietnam was occupied by Chinese forces in the north, and British forces in the south. In time, these two groups were replaced with French forces.

of whether Vietnam would remain a single, unified country or possibly three separate republics.[80]

Ho adamantly supported the existence of a united Vietnam and, as such, his objectives and those of the French leaders were diametrically opposed. In the spring of 1946 French officials established the Republic of Cochin China. This region, in particular, was viewed by the French as the key node in a reconstituted French union in Indochina. French officials also feared that if their Southeast Asian colonies broke away from French control, other, more valuable, French colonies in North Africa would likewise break away.

In the international arena of geopolitics, however, Ho had few friends when it really mattered. Although American officials were wary of the reimposition of colonialism in Southeast Asia, they were more concerned with the establishment of Communism in the region. Furthermore, the US government was more concerned with the reconstruction of Europe and the possibility of "losing" France to the socialist bloc. The Soviet Union, likewise, while supportive in general of Communist movements in Southeast Asia, was equally concerned with events in Europe, events that often overshadowed and over-determined their policies in Southeast Asia. The foremost priority for the Soviets was the promotion of communism in Europe. Joseph Stalin strongly desired a French Communist Party victory in France. However, in France the recolonization of Indochina was an popular cause, one that impelled French Communist leaders to support colonialism. Consequently, the Soviets likewise were forced to distance themselves from Ho. With no diplomatic recourse possible, the Viet Minh turned to armed conflict. On December 19, 1946 the Franco-Viet Minh (or First Indochina War) began.[81]

The First Indochina War would have tragic consequences for the people and politics of Cambodia. With the return of French authorities in late 1945, for example, Son Ngoc Thanh was arrested and sent to prison in Saigon. In 1947 he was condemned to 20 years imprisonment, a sentence later commuted to administrative surveillance in France.[82] The pliable Sihanouk, who had declared independence from the French, was left in power. Similar to the American's decision to retain Hirohito as emperor in Japan, the French well understood how most Khmers—excluding, of course, the anti-royalists—felt about the king. It was better to re-fashion a new relationship with Sihanouk rather than risk an increase in anti-French sentiment by deposing the king.[83]

Concerned primarily with the brewing conflict in Vietnam, French authorities initiated a number of reforms in Cambodia. In early 1946 a provisional agreement was reached that granted increased powers to the Cambodia government. Plans,

80 Schulzinger, *A Time for War*, 26.

81 Lasting eight years, the conflict would result in 172,708 casualties for the French and their allies; Viet Minh losses were probably three times as many. Not to be discounted are an estimated 150,000 Vietnamese civilians who died in the war. See Schulzinger, *A Time for War*, 20.

82 David P. Chandler, *Tragedy of Cambodian History: Politics, War, and Revolution Since 1945* (New Haven, CT: Yale University Press, 1991), 23–24, 26. While "detained" in France, Son Ngoc Thanh completed a law degree .

83 Tully, *Short History of Cambodia*, 111–112.

likewise, were made for an elected assembly as part of a constitutional monarchy. It was not, however, real independence since "control over military affairs and foreign relations, finances, customs and excise, posts and telegraphs and railways remained in French hands, and French officials would supervise most aspects of the Khmer government, bureaucracy and the police."[84]

Arguably the most significant of these reforms was the emergence of political pluralism. Cambodia's first ever elections were slated for April 1946 and a number of political parties emerged to contest them.[85] Many parties were short-lived and ill-formed, a not unsurprising development given the lack of any previous democratic system in Cambodia. The earliest to take shape was the Liberal Party (Kanaq Sereipheap), which was, in fact, highly conservative. Founded secretly by the French, the Liberal Party was led by Prince Norodom Norindeth, a prosperous landowner. Drawing recruiters from members and clients of the royal family, the Liberal Party retained strong ties to the monarchy, the Khmer elite, and the French.[86]

Competing against the Liberal Party were the Democrats. Founded in April, the Democratic Party (*Kanakpak procheaneathipdei*) was supported mostly by monks, teachers, students, and young civil servants. The party was led by Prince Sisowath Yuthevong, a highly educated young man—he had doctorates in both mathematics and astronomy—who had spent fifteen years living in France. Moreover, he had connections with the French Socialist Party and, being outside of Cambodia for so long, he was somewhat detached from court politics and intrigue.[87]

According to Chandler, the Democratic Party was the first political organization in Cambodia that was oriented toward the future.[88] Whereas the Liberals promised to uphold the status quo, the Democrats forwarded a platform in alignment with the French Socialist Party. They championed a modernized and democratic Cambodia, replete with universal suffrage, a bicameral parliament, and the gradual transfer of power from the French. They believed, unlike the anti-French guerrillas (see below), that independence for Cambodia could be won through peaceful, democratic processes.[89]

At stake in the 1946 elections were seats on a consultative assembly that would approve the constitution then being drafted. In preparation, the Democratic Party made a concerted effort to organize regional and provincial branches, working through Buddhist monasteries, schools, ministries, and government services. Furthermore, the party nominated candidates who commanded widespread local support. Given these strategies, the Democratic Party won a substantial victory in the elections, garnering 50 of 67 seats. The Liberal Party, conversely, won only 14 seats.

---

84 Tully, *Short History of Cambodia*, 112.

85 Tully, *Short History of Cambodia*, 113.

86 Chandler, *Tragedy of Cambodian History*, 30; see also Tully, *Short History of Cambodia*, 113.

87 Chandler, *Brother Number One*, 23.

88 Chandler, *Brother Number One*, 22.

89 Tully, *Short History of Cambodia*, 114.

Despite the semblances of democratic order, Cambodia was far from peace. On other political fronts, the many revolutions were intensifying.

Revolutionary Unities and Disunities

France's attempt to reimpose colonial dominance over Cambodia fueled the various and disparate resistance movements that were in their infancy. From the maelstrom of the Second World War, the anti-French nationalist movement was increasingly identified as *Khmer Issarak* (Free Khmer). There was, however, no single "Khmer Issarak" resistance group. Rather, throughout the 1940s various groups emerged and collapsed, formed and re-formed. The members of these various groups often held very different outlooks and perspectives. Some sought to work closely with the Viet Minh or other Communist groupings; others were adamantly opposed to Communism. Some were sympathetic to the monarchy, while others were opposed.

The coordinates of Cambodia's resistance movements of the 1940s are in part explained by the country's geography. Along the Thai-Cambodian border, but especially throughout the long-disputed provinces of Battambang and Siem Reap, the Khmer Issaraks received support not surprisingly from Thai-based organizations, including both the government of Pridi Phanomyong and the Communist Party of Thailand (CPT). As discussed earlier, after annexing Battambang and Siem Reap, the Thai promoted the notion that Cambodians in the provinces become *Khmer Issarak*, or emancipated Khmer. Prime Minister Pridi Phanomyong in particular exhibited a strong desire to rid Southeast Asia of European colonialists. Consequently, he allowed Bangkok to be a haven for independence fighters from Laos, Burma, Vietnam, Indonesia, and Cambodia.[90]

In the early years of the Khmer resistance, the strategic importance of Thailand in general, and Bangkok more specifically, cannot be underestimated. It was through the Communist Party of Thailand (CPT), for example, that many Khmer revolutionaries were first introduced to organized communism. Furthermore, from Battambang and Siem Reap, Khmer Issarak were able to launch attacks against the French, in effect creating a "second front" in the First Indochina War. Relatedly, in the face of French counter-offensives, the Khmer fighters could retreat and find sanctuary within Thailand proper. And in Bangkok, the Khmer Issarak were able to purchase food and weapons, exchange intelligence, and coordinate activities with other groups.[91]

One individual who purchased weapons in Thailand, but operated in the east, was Son Ngoc Minh. As discussed earlier, in 1942 Son Ngoc Minh, then known as Achar Mean, fled Phnom Penh following the July 20 anti-French demonstrations and established contact with the ICP in Battambang.[92] Now, using funds provided by the Viet Minh, Son Ngoc Minh bought enough materials to equip a largely Khmer company at Bay Nui across the Vietnamese border from Takeo. From this modest

---

90 Becker, *When the War was Over*, 69.

91 Becker, *When the War was Over*, 69.

92 Son Ngoc Minh, it will be recalled, was also allegedly one of the seven youths who led a coup against Sihanouk in 1945.

beginning, he established the Liberation Committee of South-East Kampuchea, which was headed by another former Khmer monk, Keo Moni.

Toward the end of the decade, attempts were made to consolidate, or at least to better coordinate, Khmer resistance activities. On February 1, 1948 the Khmer Issarak movement formed a Khmer People's Liberation Committee (KPLC; *Kana Cheat Mouta Keaha Mocchim Nokor Khmer*). According to the first manifesto of the KPLC, "French imperialism is on the wane. It will meet its death at the hands of the fraternal combat by the three peoples, Lao, Vietnamese, and Cambodia. An Indochinese Front for Independence is therefore an immediate necessity ... The Khmer people must follow the path traced by democratic and anti-colonial human-kind."[93]

The KPLC was headed by the nationalist Dap (Sergeant) Chhuon, a former member of the Cambodian militia who had deserted in 1943 and, with Thai support, had organized anti-French guerrilla bands in Battambang and Siem Reap.[94] Neither Dap Chhuon nor the Committee were Communist, although at least five of the 11 leaders of the committee—Hong Chhun, Mey Pho, Sieu Heng, Leav Keo Moni, and Muon—were sympathetic to the Vietnamese Communists and four (with the exception of Hong Chhun) were to join the Communist party.[95]

By late 1948 the Khmer resistance was divided into four military zones: the South-east, headed by Keo Moni; the North-east, under Son Sichan; the South-west, under Son Ngoc Minh; and the North-west, under Dap Chhuon as head of the KPLC. Apart from Chhuon, all zone commanders were members of the ICP and had established regional Liberation Committees closely linked to the ICP. Dap Chuon's KPLC did not have such links.[96]

In early 1949 the KPLC was reorganized as the Khmer National Liberation Committee, or KNLC (*Kana Kamathikar Khmer Sang Cheat*). A new committee was elected, with Dap Chhuon retaining the lead position. Apparently, the reorganization was an attempt to remove some of the Vietnamese Communist influence in the Committee; however, three of the other six members (Mey Pho, Leav Keo Moni and Muon) elected to the Committee were allied with the Viet Minh. Sieu Heng at this point had apparently transferred east, becoming chief of the Viet Minh's "Central Office South," with jurisdiction over Cambodia, South Vietnam, and South-Central Vietnam.[97]

Throughout 1949 Dap Chhuon's authoritarian style alienated many of the KNLC members. Leav Keo Moni (along with a hundred guerrilla fighters) and Kao Tak (also with 400 fighters), for example, each broke from the KNLC. In July Dap Chhuon was replaced by long-time veteran, Poc Khun. According to Kiernan, Poc Khun seems to have tried to unite all the various trends of the Issarak movement,

---

93 Quoted in Kiernan, "Origins of Khmer Communism," 165.

94 Chandler, *Tragedy of Cambodian History*, 33; Kiernan, *How Pol Pot Came to Power*, 58.

95 Kiernan, *How Pol Pot Came to Power*, 58.

96 Kiernan, "Origins of Khmer Communism," 165.

97 Kiernan, "Origins of Khmer Communism," 166; see also Kiernan, *How Pol Pot Came to Power*, 59.

including those sponsored by the Viet Minh. This effort, as had so many in the past, failed. In April 1950, Poc Khun was replaced by Leav Keo Moni.[98] A further blow to the Khmer independence movement occurred with the defection of Dap Chhuon. Since at least February Dap Chhuon had been seeking an accommodation with the French and in September 1949 the former guerrilla fighter, along with a force of 300, surrendered to the French. The colonial authorities quickly named him Commander-in-Chief of the "Franco-Khmer Corps."[99]

The fracturing of Khmer resistance occurred at a time that Vietnamese influence began to overshadow that of Thailand. Crucially, a rightist military coup d'état in November 1947 overthrew the Pridi government and forced the Cambodian independence groups back into their own country. Denied Thai support, many Khmer forces were forced to seek aid elsewhere. Vietnamese officials, throughout the 1940s, had repeatedly distinguished between the more urgent task of organizing military resistance to the French, and the more problematical question of organizing a Communist movement among the Khmers. However, political changes—both opportunities and obstacles—throughout Asia forced the Vietnamese leadership to revise their strategy. First, as indicated, Vietnamese Communist officials experienced on-going difficulties in cooperating with the many factions and leaders of the Khmer resistance. This accounts, in part, for the decision of the Vietnamese leadership to publically dissolve the ICP[100] in 1946 so that it would appear as if the Viet Minh united front, and not the Communists, were directing the resistance war against the French.[101] Second, Vietnamese officials were likewise concerned over their future relations with Thailand. By 1949, two years after the ouster of the Pridi government, it was clear that Phibul was establishing closer ties with the United States—based on an anti-Communist platform. Third, Vietnamese commanders were excited about the prospects developing in north Asia. In particular, the Communist victory in China and the establishment of the People's Republic of China promised a new phase of Chinese-assisted offensive actions against the French. Fourth, in 1949 the French had reached agreements with the Lao and Khmer governments vis-à-vis the Élysée Agreements. As the Franco-Viet Minh War dragged on, French officials repeatedly attempted to set up an indigenous Vietnamese regime as an alternative to Ho's DRV. On March 8 the puppet Vietnamese emperor Bao Dai and French president Vincent Auriol declared (on paper, if not in practice) that Vietnam was an independent and unified state, meaning that the disparate territories of Cochin China, Tonkin, and Annam could be treated as one entity. Concurrently, Laos and Cambodia would be "independent" states. Vietnamese Communist leaders (as well as Bao Dai himself) well understood that France's promise of independence was a chimera. French citizens would still retain special rights, and the French government would still retain ultimately control over the territories' defense, finance, and diplomatic functions. Independence, in effect, was a cover for French colonialism. However, Vietnamese

---

98 Kiernan, "Origins of Khmer Communism," 166.

99 Kiernan, "Origins of Khmer Communism," 166.

100 In fact, the ICP was not actually dissolved. It continued to operate, secretively, until 1951.

101 Becker, *When the War was Over*, 70.

officials also recognized that, discursively, the Viet Minh might be portrayed as imperialists in Laos and Cambodia: Any attempt to impose an "Indochinese" structure may present a threat to future cooperation between the Vietnamese and the many Laotian and Khmer resistance groups. Combined, these conditions impelled the Vietnamese leaders to adopt a strategy of "three fronts united" to counter France's promotion of "three independent states." Such a move by the Vietnamese aimed at demonstrating that the resistance movements in Laos and Cambodia were in fact independent entities that were voluntarily cooperating with the Viet Minh resistance movement.[102]

The "three fronts united" strategy required that each entity—Cambodia, Laos, and Vietnam—have its own party. In February 1950 the ICP called for the active construction of independent Lao and Cambodian organizations and on March 12 Vietnamese representatives, notably Le Duc Tho[103] and Nguyen Thanh Son,[104] and Cambodian representatives, including Son Ngoc Minh, Sieu Heng, Tou Samouth, and Chan Samay met at Hatien, near the Cambodian border. Throughout the ten-day conference, plans were co-ordinated for the "First National Congress of the Khmer Resistance." This congress convened on April 1950, attended by many who had not been present at the meeting in Hatien, as well as others who were not affiliated with the ICP. Overall, approximately 200 delegates assembled at Kompong Som Loeu in southwest Cambodia and at the end of the three-day meeting, the Congress established the Khmer Issarak Association (*Samakhum Khmer Issarak*). Based largely on Son Ngoc Minh's ICP-dominated Liberation Committee,[105] the Khmer Issarak Association was headed by a National Central Executive Committee, led by Son Ngoc Minh. Also included on the 15-person Executive Committee were Chan Samay, Sieu Heng, Tou Samouth, Meas Vong, and Keo Moni. Keo Meas was also apparently elected to the Executive Committee. Of the fifteen leaders, only five were definitely members of the ICP, while others were only loosely aligned with the Vietnamese organization.[106]

---

102 Porter, "Vietnamese Communist Policy," 67.

103 Le Duc Tho was a founding member of both the Indochinese Communist Party and the Viet Minh and during the French Indochina War he was chief commissar for southern Vietnam. From 1968 onward Le was the principal negotiator for the Democratic Republic of Vietnam at the Paris peace talks. He was awarded, but refused to accept, the Nobel Peace Prize.

104 At the time, Nguyen Thanh Son was commander of Viet Minh forces in Cambodia and chief of the Viet Minh's committee for foreign affairs in southern Vietnam.

105 Kiernan, "Origins of Khmer Communism," 167; elsewhere, the Khmer Issarak Association is referred to as the "Unified Issarak Front," which, according to Kiernan, is the incorrect French official translation of the Khmer title; see Kiernan, *The Pol Pot Regime*, 13.

106 Kiernan, *How Pol Pot Came to Power*, 79. Initially, the KNLC remained separate from the Khmer Issarak Association. Following a delegation attended by KNLC representatives Muon, Mey Pho, and Hong Chhun, the KNLC joined with the Khmer Issarak Association in spring 1951. Many other Khmer Issarak groups refused to cooperate with the Viet Minh. In Kompong Speu, for example, both Prince Chantaraingsey and bandit leader Puth Chhay led militias of over 1,000 troops. In 1953 these two groups would rally to the French. Son Ngoc Thanh, who by this point had returned to Cambodia, also refused to align with the Viet Minh.

The April 17 Congress likewise established a "proto-government," known as the People's Liberation Central Committee (PLCC; *Kanak Mouta Keaha Moçchim Ban Dos Asan*). Also headed by Son Ngoc Minh, the PLCC included three vice-presidents, each of which represented one of the new zones established for Cambodia: Chan Samay, in the southwest (under Son Ngoc Minh); Sieu Heng, in the northwest; and Tou Samouth, in the east (under Keo Moni). According to Kiernan, this organization may well have been the effective governing body of the Khmer Issarak Association. Indeed, it was under the auspices of the PLCC that Son Ngoc Minh would declare Cambodia's independence on June 19, 1950.[107] His government was immediately recognized by the Viet Minh and by the Lao resistance movement.

The early 1950s were pivotal years in the development of the Cambodian revolution. First, Son Ngoc Minh emerged as a key leader of the Khmer liberation movement. By working within established structures (i.e., the earlier Khmer Issarak Committee), and through an emphasis on training Khmer cadres in party schools, Son Ngoc Minh substantially broadened participation in the movement. Furthermore, throughout the 1950s Son Ngoc Minh, working with Nguyen Thanh Son, established in Military Zones political-military schools. Classes were taught by Vietnamese cadre, but in the Khmer language, with the goal of training training hundreds of Khmer in party schools. Son Ngoc Minh thus was able to initiate a rapid expansion of the movement, in close cooperation with the Vietnamese Communists, but he was unable to overcome one major limitation. The vast majority of KPRP members were ethnic Vietnamese living in Cambodia; the party itself still had minimal appeal to indigenous Khmers, many of whom worried that communism was the means of furthering Vietnamese dominance.[108]

Outside of Cambodia, in early 1951 the ICP was formally dissolved and renamed the Vietnamese Workers' Party (VWP, Dang Lao dong Viet Nam). This move coincided with Vietnamese efforts to facilitate the formation of separate revolutionary parties and by September 1951 the Khmer People's Revolutionary Party (KPRP) was officially founded. The KPRP was led by a provisional central committee of 15 members. Son Ngoc Minh was once again elected to lead the committee; other members included Tou Samouth (deputy), Sieu Heng, and Chan Samay. Other committee members apparently included former Khmer ICP members, such as Mey Pho, Keo Moni, Keo Mas, and So Phim; others were ethnic Vietnamese.[109]

The KRPR was heavily influenced by the Vietnamese. The final draft of the KRPR's statutes, written in Vietnamese and translated into Khmer, consisted of a "simplified version" of the VWP statutes. Tellingly, the Party's mandate was only to fight imperialism and not to work for socialism. Indeed, the statutes contained no mention of Marx or Lenin. Such moves were strategic. Aware of the distrust exhibited by many Khmer vis-à-vis Vietnam and Communism, the long-term Communist agenda of the movement was deliberately down-played. Cambodians

---

None of these groups possessed a nation-wide organizational structure and many fought not only among themselves, but against the Viet Minh *and* the Khmer Issarak Association. See Kiernan, "Origins of Khmer Communism," 170.

107 Kiernan, *How Pol Pot Came to Power*, 80.
108 Kiernan, "Origins of Khmer Communism," 168.
109 Kiernan, "Origins of Khmer Communism," 169.

who were recruited to the KRPR were often told that its aims were to drive the French from Indochina and to cooperate with the Vietnamese toward this goal.[110]

It was at this point that a number of Khmer students, fresh from their studies abroad in France, began to participate in the Khmer liberation movement. Notable among these young students was a secretive man known as Saloth Sar.

On May 25, 1925 Saloth Sar was born in the village of Prek Svauv, near the provincial capital of Kompong Thom, some ninety miles north of Phnom Penh.[111] His father was a prosperous farmer with nine hectares of land, several draft cattle, and a comfortable tile-roofed house. He was the eighth of nine children.[112] Not of peasant background, Saloth Sar never worked a rice field or knew much of village life.[113] In fact, his family had royal connections. In the 1920s, as King Sisowath's (r. 1904–1927) reign was nearing its end, Saloth Sar's cousin, Luk Khun Meak, joined the royal ballet. She soon became a consort of the king's eldest son, Prince Sisowath Monivong. In 1927, when Monivong ascended to the throne, Meak held the favored position of *khun preah me neang* or, "lady in charge of the women." In addition, Sar's older brother, Loth Suong, worked at the palace beginning in the late 1920s as a clerk. Their sister, Saloth Roeung, joined the ballet and in the 1930s she also became a consort of King Monivong. Sometime around 1934 or 1935 Saloth Sar and an older brother, Chhay, went to live with Meak and Suong in Phnom Penh. Sar spent several months at Vat Botum Vaddei, a Buddhist monastery. It was here he learned the basics of Buddhism and became literate in Khmer.[114] Later he attended École Miche, a private Catholic school.

In 1942 Saloth Sar studied at the Collège Norodom Sihanouk in Kompong Cham. He left sometime around 1947, and in 1948 he enrolled as a carpentry student at the École Technique at Russey Keo. It is unclear if Saloth Sar failed to pass the examinations that would have allowed him to pursue a baccalaureate at the more prestigious institution, Lycée Sisowath, or if he in fact wanted to obtain a technical degree. Irregardless of the reason, he did not stay long, for in 1948 he received a government scholarship to study radio-electronics in France.[115]

Saloth Sar was one of a select group of Cambodians, those few students who were fortunate enough to obtain higher education. And it was in France that Saloth Sar began to develop and cultivate his more radical ideas. In was in Paris, for example, that Saloth Sar adopted his *nom de plume*, Pol Pot, which means "the Original Cambodian."[116] It was also during this time that Pol Pot became interested in Marxism and became involved in the Cambodian section of the French Communist Party.

---

110 Chandler, *Brother Number One*, 31.

111 Saloth Sar's birth is the subject of intense debate. Different accounts indicate either May 19 or May 25; likewise, some accounts use 1925 while others use 1928. In part, confusion reigns because Saloth Sar himself was so secretive as to his personal life.

112 Chandler, *Brother Number One*, 8.

113 Kiernan, *The Pol Pot Regime*, 9.

114 Chandler, *Brother Number One*, 8–9; see also Kiernan, *How Pol Pot Came to Power*, 26–29.

115 Chandler, *Brother Number One*, 21.

116 Kiernan, *The Pol Pot Regime*.

Pol Pot befriended a number of other like-minded Cambodians in Paris, many of whom would remain his closest allies for the next four decades. One such man was Ieng Sary.[117] Sary would ultimately emerge as the Minister of Foreign Affairs of the Democratic Republic of Kampuchea. Saloth Sar, while in Paris, also met Khieu Ponnary, a woman eight years his senior. She would become the first Khmer woman to earn a baccalaureate; she also would become Pol Pot's first wife.

A close-knit group was developing. Throughout the ensuing years, Sary would marry Khieu Ponnary's sister, Khieu Thirith. Like her sister, Khieu Thirith was highly educated. In Cambodia she had graduated from the prestigious *Lycée Sisowath* and, in Paris, she majored in Shakespearean studies at the Sorbonne. In the process, she became the first Cambodian to achieve a degree in English literature. Years later Thirith would become a senior member among the Khmer Rouge leadership, serving as Minister of Social Affairs and Head of Democratic Kampuchea's Red Cross Society. Other members of the Paris-educated elite would include Khieu Samphan, Hou Youn, and Son Sen.

Saloth Sar interacted with many other Khmer students, including the radicals Thiounn Mumm and Keng Vannsak. Many of these students came from the highest echelons of Cambodian society. Thiounn Mumm, for example, was a member of the most powerful non-royal family in Phnom Penh. His grandfather had amassed a fortune as minister of palace affairs, and his father was the first Cambodian to earn lycée and university qualification in France. Mumm, moreover, had studied at the University of Hanoi and, in 1946, was one of the first Cambodian students to travel to France for higher education. Specializing in applied science, Thiounn Mumm most likely introduced Saloth Sar to the Communist party.[118] Keng Vannsak likewise had royal connections and had attended Lycée Sisowath. He was sent to France to study Cambodian linguistics. Although drawn to the Left, Keng Vannsak never became a Communist; he was, however, an ardent nationalist with anti-royalist leanings.[119]

As will be discussed in greater detail in Chapter 4, Saloth Sar and his peers began to develop their revolutionary ideas while in Paris. In particular, through various reading groups, these students read, learned, and debated the meanings of revolution, the strategies involved, and the objectives. Increasingly, these revolutionary ideas were couched in extremely nationalistic terms. The revolution was to be completely autonomous, with the objective of creating a self-determined and sovereign Cambodia. Such a direction would ultimately set these Paris-educated revolutionaries on a collision course with the Vietnamese-influenced Khmer revolutionaries.

---

117 Chandler, *Brother Number One*, 22. Born as Kim Trang, the son of mixed Khmer-Chinese ancestry he was raised among the Cambodian minority in Cochin China. And like Saloth Sar, Kim Trang did not come from peasant roots; indeed, his father was a prosperous landowner. However, Trang's father died and he was sent to live with relatives, who promptly renamed him Ieng Sary. To his nationalist-minded adoptive parents, "Ieng Sary" sounded more Khmer.

118 Chandler, *Brother Number One*, 27.

119 Chandler, *Brother Number One*, 27.

The Move Toward Independence

In 1951 Son Ngoc Thanh returned to Cambodia. The impetus of his return was most likely pressure placed on Sihanouk by the King's father and his father's friends in the Democratic Party. Sihanouk no doubt held out hope—much like the Paris-educated Khmer who visited Thanh in France—that the nationalist could be politically useful. For Sihanouk, Thanh represented the possibility of splitting the Democrats; for the Paris-educated students, Thanh represented the ascension of the Leftist-oriented independence movement. Both sides were to be disappointed.[120]

Son Ngoc Thanh was politically inactive throughout 1951. He refused several cabinet posts, including that of minister of foreign affairs, claiming that he lacked ambition. In December, according to an American diplomat, Thanh explained that he had been outside of Cambodia for a long while and wanted to continue conversations with all elements before deciding on a course of action.[121] In early 1952, however, he began touring the provinces with his long-time friend and recently named minister of information, Pach Chhoeun. During the tour, Thanh played down the importance of Sihanouk, angering the king. Sihanouk was convinced the Americans were using Thanh toward their own ends. Around February Thanh founded a newspaper, *Khmer Krok* ("Cambodians Awake"). He explained in the first issue: "We know that the Cambodian people, who have been anaesthetized for a long time, are now awake … No obstacle can now stop this awakening from moving ahead."[122]

On March 9, 1952, the 7th anniversary of the Japanese coup, Son Ngoc Thanh disappeared. Thanh, a leftist intellectual named Ea Sichau, and a handful of others left Phnom Penh. Their destination was supposedly a customs facilities on the Thai border. By nightfall, they had not arrived. It was assumed that Thanh had been kidnapped, perhaps by Issarak guerrillas. In actuality, Thanh had set up his own political base along the Thai border in northern Siem Reap province. Thanh apparently hoped to join forces with non-communist, anti-French Issaraks and to win over pro-Communist guerrillas. However, Thanh misjudged his popularity and only a handful of people followed. Son Ngoc Thanh had gambled and failed on his ability to unite an anti-French resistance movement. From 1952 onwards, he became marginal to the larger independence struggle.[123]

The defection of Thanh was unsettling for all concerned groups. Sihanouk and his French supporters believed that the Democrats had sponsored his defection, a thesis believed also by conservatives. French officials, however, also considered the possibility of removing Sihanouk. Within this context, Sihanouk dismissed (in June 1952) the newly-elected and Democrat-dominated cabinet and dissolved the National Assembly. He then proceeded to name a new, non-Democratic cabinet and took over the position of prime minister; his cousin, Sirik Matak, was placed in

---

120 Chandler, *Tragedy of Cambodian History*, 57.

121 Chandler, *Tragedy of Cambodian History*, 58.

122 Chandler, *History of Cambodia*, 180.

123 Chandler, *Brother Number One*, 36; Chandler, *History of Cambodia*, 180. During the next few years Son Ngoc Thanh continued his efforts to influence political affairs in Cambodia. He aligned himself with the United States, Thailand, and South Vietnam. He proved to be a bothersome thorn in Sihanouk's side, constituting a potential, albeit distant, threat.

charge of defense. Sihanouk's coup marked the end of the Democrat's participation and closed an era of political pluralism. According to Chandler, the "king was no longer willing to play the role of a constitutional monarch or to be an instrument of French colonial policy." For the next 18 years Sihanouk would govern directly or through proxies.[124]

The coup itself was relatively peaceful. Demonstrations, however, soon followed, particularly among Cambodia's lycées, where anti-monarchic, pro-Democrat sentiment ran high; radical students in France, likewise, condemned Sihanouk as a traitor to their country. Sihanouk, countered, however, with the announcement that he would deliver independence within three years.[125]

In February 1953 Sihanouk traveled to France for "health" reasons. Over the years, in fact, Sihanouk regularly went to France. This time, however, he departed on what would later be known as the "royal crusade" for independence. In France, he argued that he alone could establish a neutral and independent Cambodia. Furthermore, he would forbid the communists' use of Cambodian territory in their fight against the French. However, the French people—but most importantly the French government— were nonplussed by Sihanouk's actions. Rebuffed in France, Sihanouk then traveled to Canada, the United States and Japan, seeking any sympathetic ear he might find. In the United States, however, Secretary of State John Foster Dulles informed the Cambodian monarch that independence was meaningless without French support, warning that Cambodia would be "swallowed" by the Communists. Officials in Japan, likewise, counseled Sihanouk to cooperate with the French.[126]

When Sihanouk returned to Cambodia, he placed himself in a self-imposed "exile" in the northwestern province of Battambang. In France, meanwhile, opinions were gradually changing. The war in Vietnam was going badly for the French, and it had become increasingly unpopular at home. Added to this, Sihanouk's resistance to the French and the prospect of the Indochina War expanding drastically into Cambodia worried French officials. Consequently, in October 1953 the French granted the king authority over Cambodia's armed forces and judiciary and foreign affairs. These were limited gains—the French still retained an economic hold on the kingdom, in particularly over the rubber plantations. Sihanouk therefore upped the ante by organizing mass demonstrations in his favor.[127] Bogged down with the war in Vietnam, and only marginally concerned with Cambodia, on November 9, 1953 Paris acquiesced. Sihanouk proudly declared himself the father of Cambodian independence.[128]

John Tully explains that after the success of his Royal Crusade for Independence, Sihanouk was confronted with two interlinked challenges. First, as a would-be absolute ruler, he had to consolidate his grip on domestic power, and second, he had

124 Tully, *Short History of Cambodia*, 119; Chandler, *Tragedy of Cambodia*, 63.
125 Chandler, *History of Cambodia*, 184.
126 Chandler, *Tragedy of Cambodia*, 68.
127 Chandler, *Tragedy of Cambodia*, 68–72; Chandler, *History of Cambodia*, 185–187.
128 Becker, *When the War was Over*, 76.

to maintain Cambodia's sovereignty in the uncertain world of the Cold War. [129] His first mile-post occurred at the Gevena Conference.

The Geneva Conference opened on May 8, 1954 with delegations from the United States, France, the United Kingdom, the Soviet Union (USSR), the People's Republic of China (PRC), Cambodia, Laos, the Democratic Republic of Vietnam (DRV), and the Republic of Vietnam (RVN). The Conference had a dual purpose: to negotiate settlements of the Indochina War and the Korean War. Cambodia was marginal to the conference. The Khmer communist movement, in particular, was sacrificed, as both the USSR and China made concessions to the Western bloc.

The Cambodian delegation consisted of three individuals appointed by Sihanouk: Nong Kimny, Sam Sary, and Tep Phan. The Khmer Issarak Association also *attempted* to seat their own representatives, Keo Moni and Mey Pho, at the Conference. The DRV's Pham Van Dong, backed by the PRC and the USSR, proposed that representatives of the communist-dominated "resistance movements" from Laos and Cambodia be seated at the Conference. However, the Sihanouk representatives voiced strenuously their objections to such a move. Subsequently, both the PRC and the USSR backed down and persuaded Pham Van Dong to drop the demand. China in particular wanted peace on its southern border and to establish a zone of security. They were content to have a relatively calm and secure northern Indochina (thus sacrificing the communist movement in Cambodia).[130] Consequently, the Khmer representatives were not recognized as belligerents, not awarded official status, and not allowed to make public statements. Cambodia supposedly was independent.

Furthermore, Conference participants decided that Khmer resistance forces, unlike those in Vietnam and Laos, should not be given control of any national territory. They were to lay down their arms, and national elections to be held in 1955. And Cambodia retained right to enter into military alliances with other countries.

At Geneva, the Viet Minh proposed a temporary division at the thirteenth parallel, just north of Saigon, with elections to determine national unity to be held in six months. The French countered with a boundary along the eighteenth parallel, just south of Hanoi, with no elections. As a compromise, Soviet foreign minister Vyacheslav Molotov proposed the seventeenth parallel, a location just north of Hue. Elections were to be held, supervised by international observers, in two years. Dismayed by the course of events, Ho Chi Minh appealed personally to the Chinese foreign minister, Zhou Enlai. However, the Chinese leadership encouraged Ho to accept the settlement.[131] Chinese officials were concerned primarily about further American intervention in Asia, as well as the possibility for greater security arrangements between the PRC and the United States.[132]

---

129 Tully, *Short History of Cambodia*, 124.

130 Tully, *Short History of Cambodia*, 126.

131 Neale, *A People's History*, 33.

132 There exist other reasons why the PRC apparently "failed" to support Vietnam at the Geneva Convention. Chinese Communist leaders, for example, may have desired to prevent the formation of a strong regional power (Vietnam) on China's southern border. Also, China apparently wanted to be seen as separate from the USSR.

Following the Soviet's proposal, two military zones were established in Vietnam. To the north was the DRV and to the south, the State of Vietnam. Among the various conditions of the accords was a cease-fire throughout Vietnam, to be accompanied by troop withdrawals of French forces in the north and Viet Minh forces in the south. National elections were to be held in 1956.

With respect to Cambodia, the agreement stipulated that all Viet Minh military forces be withdrawn within 90 days and that Cambodian resistance forces be demobilized within 30 days. In a separate agreement signed by the Cambodian representative, Sam Sary, the French and the Viet Minh agreed to withdraw all forces from Cambodia by October 1954.[133] Elections for a reconstituted National Assembly, overseen by international observers from Poland, India, and Canada, were to take place the following year. And perhaps most significantly, the Cambodian Communist representatives were allowed no territorial stake in the country after the conference.[134]

Few parties were pleased with the outcomes of the conference. The Cambodian Communists felt betrayed by both the PRC and the DRV, while the Vietnamese Communists felt betrayed by the PRC. These feelings of mis-trust and betrayal would fester in the years ahead.

Post-Liberation Prospects

For most Cambodians, the war was over. Cambodia was independent and all foreign troops—the French colonial army and the Viet Minh—agreed to leave under the Geneva Accords. Free elections, moreover, were forthcoming in 1955.

Given these developments, many Khmer Issaraks and other resistance fights laid down their arms and went back to their villages and farms.

The Cambodian Communists faced two monumental challenges. First, they were left with no clear platform.[135] Based on Sihanouk's crusade in 1953, and the subsequent Geneva Convention, Cambodia was for all intents and purposes an independent state. Liberation was thus no longer a rallying cry for the revolutionaries. To be sure there remained strong elements of anti-monarchy sentiment, but for many Khmer citizens, Sihanouk was the "father" of independence. Who else but Sihanouk stood up to the French and delivered independence? Nor could they demand democracy—not with elections scheduled for the following year. Equally troubling was the fact that the Cambodian revolutionaries had no territory in which they could safely remain and build their movement. In short the Communist movement was falling apart; the Khmer party had no platform, no population under its control, no arms to protect itself, and no legal rights.[136]

Within the aftermath of the Geneva Conference, and faced with a failing revolution in Cambodia, Communist leaders in Vietnam made two fateful decisions. First, the Vietnamese refigured the role of Cambodia within the overall struggle for

---

133 Seekins, "Historical Setting," 28.
134 Chandler, *Tragedy of Cambodian History*, 73.
135 The changing "goals" of the revolution are considered in greater detail in Chapter 4.
136 Porter, "Vietnamese Communist Policy," 71.

independence. Vietnam, based on the Geneva Accords, remained a divided country. *Their* goal of a united, and sovereign communist state remained unattained. And it was apparent that the United States would replace the role of the French as barrier to their objective. Consequently, the Vietnamese Communists looked to the 1955 Cambodian elections as the best means of advancing its primary postwar interest in Cambodia: blocking US military penetration of the country.

The Vietnamese Communists looked to the Democrat Party, which had long opposed any Cambodian alignment with the United States, as the most viable strategy for Cambodia.[137] And in fact, events in Cambodia appeared to work in Vietnam's favor. Even before the Geneva Convention Keng Vannsak and Thiounn Mumm (both having returned from France), and several others, pushed the Democrat Party into a more hard-line anti-royalist and anti-American position. In conjunction with the activities of other Paris-educated revolutionaries, including Saloth Sar, the newly transformed Democrats hoped to foment change within the legal political system.

A second decision concerned the actual members of the Khmer Communist movement. In an effort to protect the embattled KRPR, the Vietnamese Communists split the Khmer movement into two components. One group, numbering around 1,000, would travel to North Vietnam to receive military and political training. Included among these members was Son Ngoc Minh, still the highest ranking member of the KRPR, and most of the experienced and time-tested veterans of the anti-colonial movement: Sieu Heng (who would return within months), Mey Pho, So Phim, Keo Moni, Leav Keo Moni, Chan Samay[138] Apart from protecting the KPRP, this move held the added advantage—or so it was thought—of the Vietnamese being able to dictate the activities of the Cambodian revolution in order to further Vietnam's struggle.

A second group of Khmer revolutionaries would remain behind in Cambodia. Some, such as Keo Meas, would form a legitimate political party, known as the *Pracheachon* (People's Group), to contest the 1955 elections. Others, like Tou Samouth, Nuon Chea, and So Phim, would obtain influential positions in Cambodian society, such as teachers, writers, or journalists, but work "underground" for the movement. In this way the Khmer Communists would be able to disseminate party information and to recruit new members.[139] It was at this moment that the Paris-educated revolutionaries began to make inroads into the Cambodian Communist movement.

Geographically, the KPRP was divided into two broad regions. Although Son Ngoc Minh remained the titular head of the Khmer Communists, Sieu Heng was placed in control of the Cambodian-based movement. More specifically, he oversaw political activities in the rural areas. Tou Samouth, the third-ranking official behind Son Ngoc Minh and Sieu Heng, was responsible for party affairs in the capital and other provincial towns. This geographic realignment of the communist movement in Cambodia set in place a drastic reconfiguration, one that would facilitate the emergence of the Paris-educated Cambodians.

137 Porter, "Vietnamese Communist Policy," 70–71.

138 Kiernan, "Origins of Khmer Communism," 175.

139 Becker, *When the War was Over*, 80.

For his part, Sihanouk saw no reason to share power. He believed that he had singlehandedly won Independence and, consequently, sought to play a larger role in Cambodian politics. Indeed, in a continuation of his 1952 coup, Sihanouk planned to govern his kingdom unopposed. Accordingly, he viewed the scheduled elections as both obstacle and opportunity.[110]

The 1955 elections offered Cambodian voters a wide range of political options. As summarized by Chandler, on the left were both the concealed Indochina Communist Party and its front group, the Pracheachon, as well as the radical Democrats, led by Thiounn Mumm and Keng Vannsak. In the center were Thanhist Democrats and other members of that party who were frightened by the radical inroads made by the newcomers. And toward the right were those affiliated with the Liberal Party and other anti-democratic groups formed in the early 1950s.[141]

Sihanouk was not to subject his political future to something as malleable as an election. Most observers projected a Democrat victory, given that party's organizational skills and popularity. As such, Sihanouk embarked on a campaign of brutal repression and political subterfuge. In February 1955 he sponsored a referendum asking voters to approve his Royal Crusade for Independence. Although balloting was open, Cambodian citizens who wished to vote against the king were required to discard a ballot with his picture on it, an act of lèse-majesté that was probably grounds for arrest. Not surprisingly, 95 percent of the electorate voted to support Sihanouk. Taking this result as his cue, Sihanouk then manuevered himself into a position to more explicitly participate in national politics. According to Cambodia's constitution, the position of king was primarily ceremonial. Consequently, Sihanouk made the remarkable decision to abdicate his throne, naming his father, Prince Suramarit, king.[142] According to Tully, aside from permitting Sihanouk to enter directly the political arena, this move also preserved the monarchy, which Sihanouk believed to be the social cement of the nation.[143]

Ideologically, Sihanouk moved to preempt the foreign policy positions advanced by the Democrats and the Pracheachon. In April Sihanouk attended the Bandung Conference in Indonesia and, upon his return, pledged to pursue a non-aligned policy. However, to counter any potential conservative opposition, Sihanouk also signed a military aid agreement with the United States. In effect, Sihanouk's "neutralist" position was a balancing act, designed to neutralize opposition from both sides of the political spectrum. His most decisive move, however, occurred in mid-1955 with the formation of his own political movement, the *Sangkum Reastr Niyum* ("Popular Socialist Community"). This movement was to be "the political umbrella for the whole nation and Cambodia became a de facto one-party state in which civil servants were expected to join the Sangkum."[144] Those who refused to join the Sangkum, and in particular those Democrats working in the government, were fired from their jobs.

140 Chandler, *Brother Number One*, 44.
141 Chandler, *Brother Number One*, 46.
142 Chandler, *Brother Number One*, 47.
143 Tully, *Short History of Cambodia*, 129.
144 Tully, *Short History of Cambodia*, 129.

Concurrently, those working for other political parties were harassed, beaten, and arrested. Radical newspapers were shut down, and their editors imprisoned.[145]

As the elections neared, the repression intensified. Democratic rallies were broken up and candidates were beaten. On the eve of the elections Keng Vannsak was arrested and held without trial for several months; Thiounn Mumm fled to France. And if these physical acts of violence were not sufficient, Sihanouk resorted to outright fraud and corruption. In the end, Sangkum candidates "won" all 91 seats and 630,000 votes—over 80 percent of those cast. As Chandler summarizes, the numbers convinced Sihanouk that Cambodia could now be managed as his own personal estate.[146]

Following the 1955 elections Sihanouk worked to secure his authority over Cambodia. Sensing the greatest threat from the small, but ever-present communist movement, Sihanouk pursued a non-declared—and largely covert—campaign against members of the communist party. Through assassination and intimidation, bribery and cajolery, Sihanouk moved to both co-opt the appeal of the left as well as to eradicate individual members of the Left. On the other hand, to neutralize anti-imperialist rhetoric, and to secure support from communist powers (North Vietnam and the PRC), Sihanouk—although not a communist—embarked on a leftist reorientation of the economy.

Internationally, Sihanouk assumed that the US would ultimately lose in Vietnam and that China would become the dominant external influence in Southeast Asia. Sihanouk, though, was a pragmatist and a realist. Despite his overtures to China, he continued also to seek aid from the United States, so much so, in fact, that by the early 1960s aid from Washington constituted 30 percent of Cambodia's defense budget and 14 percent of its total budget. Between 1955 and 1963 economic and military aid from the United States totaled approximately US \$400 million.[147] However, given the past history of Cambodia, including the dominance of both Thailand and Vietnam, Sihanouk was less than eager to enter into any alliance that included these neighboring states. Thus, Sihanouk was reticent to the idea of an American-based alliance that included both Thailand and the fledgling state of South Vietnam.

The Revolution at Decades' End

By 1959 Cambodian politics were dominated by Sihanouk and his Sangkum movement. The Khmer Communist movement was geographically and ideologically fragmented. Most party veterans were living in exile in North Vietnam; those who remained in Cambodia were forced to work underground, or risk harassment (or assassination) by Sihanouk's forces.

To compound problems, Vietnamese Communist leaders continued to forestall the possibility of armed insurrection in Cambodia. From the Vietnamese point of view, the on-going, and intensifying, struggle against the United States was paramount. It was imperative that Cambodia remain neutral so as to prevent the United States

---

145 Chandler, *Brother Number One*, 46–47.

146 Chandler, *Brother Number One*, 49.

147 Seekins, "Historical Setting," 32–34.

from establishing a base of operations on Vietnam's western border. As such, the Vietnamese party urged the Khmer to unite with Sihanouk in his opposition to US intervention, while waging peaceful political struggle—through the Democrat Party and the Pracheachon—against Sihanouk's domestic policies. The Khmer Communists were thus placed in the disastrous and self-defeating position of publically supporting Sihanouk, all the while Sihanouk was conducted a terror campaign to rid himself of the communist threat. Some Cambodian members, including Tou Samouth, apparently accepted the Vietnamese strategy; others, including the Paris-educated Khmer Communists, were adamantly opposed.[148]

And in 1959, the prospects for a Communist revolution in Cambodia further dimmed with the defection of Sieu Heng, the second-ranking official of the Party. After independence had been achieved in 1954, Sieu Heng apparently saw little reason for revolution. Rather than simply walking away, however, Sieu Heng remained active, but beginning in 1955 he began to work as an informant for Sihanouk's then police chief, Lon Nol. During the next four years, until his public defection in 1959, Sieu Heng provided the government with the names of Khmer Communists living and working in the countryside. Lon Nol, in turn, set about eradicating the rural cells. By 1959 the Khmer Communist movement had lost nearly 90 percent of its rural cadre. Some were murdered, some quit the movement out of fear of retribution. Some simply disappeared.[149]

A geographic consequence of Sieu Heng's defection was that the rural stronghold of the Khmer communists collapsed. Only in the cities, and especially Phnom Penh, did the Khmer Communist movement continue. And it was in the cities that Pol Pot and the other Paris-educated Khmer were in control.

---

148 Porter, "Vietnamese Communist Policy," 73.

149 Becker, *When the War was Over*, 81. In the end, Sieu Heng paid for his betrayal. In 1975, after the Khmer Rouge came to power, Nuon Chea, who was in fact married to Sieu Heng's niece, enticed the former communist, by then half-paralyzed, out of retirement with promises of rewards as "father of the revolution." Sieu Heng was put to death. See Chandler, *Tragedy of Cambodian History*, 33.

Chapter 3

The Improbable Revolution

Throughout most rural areas of Cambodia in the late 1950s, preconditions for revolutionary action were difficult to discern.[1] Sihanouk's crusade for independence had obviated, in the minds of many Khmer, the need for revolution. Forrest Colburn notes that often, many "ordinary" or rank-and-file members of revolutions fight not *for* something, but rather *against* something.[2] In Cambodia, many Khmer were not fighting for a particular ideology, rather, their revolution was anti-colonial, and hence *against* colonial domination. Most were decidedly not fighting *for* communism. Consequently, as David Chandler explains, most Cambodians "were reluctant to become involved in rebellious politics after Cambodia's independence had been won."[3] Moreover, the individualism, conservatism, Buddhist ethics, and the fact that nearly all of the Cambodian peasantry actually owned their land made them unlikely candidates for Communist recruitment.[4]

Even had conditions been optimal, the Khmer communist movement would have been in no position to progress. In practical terms, the remnants of the communist movement were in disarray—weakened by both the defection of Sieu Heng and of Sihanouk's on-going (albeit tempered) repression. The Khmer communists had neither a territorial base under its control, nor a substantial population supporting its actions. It had no arms to protect itself and no legal existence. In fact, in 1969 the Khmer Rouge probably numbered no more than 4,000. And following the humiliation of the Geneva Conference, the fledgling Cambodian communist movement apparently had no foreign benefactors either. Neither the Democratic Republic of Vietnam (DRV), People's Republic of China (PRC), nor the Soviet Union (USSR) seemed prepared to support their communist brethren in Cambodia. Indeed, these states were more apt to sacrifice the Khmer to further other, supposedly more important objectives. The DRV, in fact, actually encouraged the Khmer to align themselves—at least publically—with Sihanouk and to *not* wage armed struggled against the monarchy. Sihanouk, for his part, continued his repression of the Khmer communists, while at the same time moving his policies, ironically, closer to those of the DRV and

1 David P. Chandler, *The Tragedy of Cambodian History: Politics, War, and Revolution Since 1945* (New Haven, CT: Yale University Press, 1991), 108.
2 Forrest D. Colburn, *The Vogue of Revolution in Poor Countries* (Princeton, NJ: Princeton University Press, 1994), 46.
3 Chandler, *Tragedy of Cambodian History*, 108.
4 Chandler, *Tragedy of Cambodian History*, 108.

PRC. As Elizabeth Becker concludes, "That the communists survived this period was somewhat miraculous."[5]

The Khmer communist movement did survive, however, and would "unexpectedly" achieve victory within twenty years.[6] That this occurred owes as much to the brutality of the Khmer Rouge leadership as it does to external events that literally forced many Cambodians to support the communist movement.

David Chandler explains that the Khmer People's Revolutionary Party's (KPRP) mission after 1955 was both simpler and impossible. It was easy, for example (in the minds of the KPRP) to identify the enemy and to outline the conditions for revolution. The Cambodian people, it was argued, were exploited by merchants, moneylenders, corrupt officials, the royal family, pro-US elements, and so on. Furthermore, the Khmer Rouge argued that Cambodia, in fact, was not independent. Rather, it was a semi-colonial state, dominated by feudal landowners, reactionary classes, and foreign governments. This was the message that the movement attempted to promote, a dual anti-royalist and anti-imperialist platform. All that was required was a proper education for the masses, an intense training period that would be led by a vanguard dedicated to the liberation of the peasantry.

Such a task, however, proved daunting. It was difficult, for example, to portray Sihanouk as a puppet of the United States. The monarch forwarded an anti-American stance; he installed known Leftist politicians into the Sangkum, including high-level cabinet positions;[7] he allowed the Pracheachon to function as political group (though under surveillance and susceptible to terrorist harassment);[8] and beginning in 1956, he embarked on a program of "Khmer socialism." To make matters more difficult, the Cambodian people continued to see Sihanouk as the father of independence.[9]

Sihanouk, of course, was not actually moving toward a Communist position. He was, in fact, performing a balancing act, a performance intended to serve several

---

5 Elizabeth Becker, *When the War was Over: Cambodia and the Khmer Rouge Revolution* (New York: Public Affairs, 1998), 80; see also William Shawcross, *Sideshow: Kissinger, Nixon, and the Destruction of Cambodia*, revised edition (New York: Cooper Square Press, 2002), 73.

6 Timothy Carney, "The Unexpected Victory," in *Cambodia 1975–1978: Rendezvous with Death*, Karl D. Jackson (ed.) (Princeton, NJ: Princeton University Press, 1989), 13–35.

7 In 1958, for example, Sihanouk appointed the communists Hou Youn and Hu Nim to his cabinet and in 1961 he appointed Khieu Samphan and Hou Youn as secretary of state for commerce and secretary of state for planning, respectively. Through these latter two appointments, for example, Sihanouk was able to co-opt two potentially errant subjects. At the same time, however, he could take advantage of two talented economists who if successful, could stabilize the economy, for which Sihanouk could take credit. If the policies of Khieu Samphan and Hou Youn failed, of course, Sihanouk could blame the Left-leaning appointees and use their failure as propaganda against the Communist movement.

8 For example, Sihanouk allegedly arranged for the editor of *Pracheachon's* newspaper to be murdered. Philip Short explains that Sihanouk's campaign of repression reduced the number of Party members by half, from 1,670 at the end of the war to 850 by 1957; most of the former rural leaders, moreover, were inactive. Many, such as Ruos Nhim, Ke Pauk, and So Phim, were working as either farmers or carpenters. See Philip Short, *Pol Pot: Anatomy of a Nightmare* (New York: Henry Holt and Company, 2004), 120.

9 Chandler, *Tragedy of Cambodian History*, 108.

objectives. First, Sihanouk sought to consolidate his own power and to remain—as had the Angkorian kings of the past—the absolute ruler of his land. The death in 1960 of Sihanouk's father, King Norodom Suramarit, for example, posed both problems and opportunities. Sihanouk well understood that the monarchy was the social glue that bound the nation together. If he were to abolish the monarchy—which, in fact, was a goal of the Communist movement—Sihanouk feared that he would lose control of his country. However, his own personal power might be severely diminished if a strong successor to the throne emerged. And there were many possible heirs. Sihanouk contemplated placing his mother, Queen Kossamak, or his uncle, Prince Sisowath Monireth, on the throne. Sihanouk's eldest son, Prince Rannaridh, was not yet of age, and thus was not an option. There were, however, perhaps one hundred other princes that might be considered. All of these options posed considerable risks for Sihanouk. His solution was both ingenious and self-serving. Sihanouk would remain as head-of-state (but not King) and Monireth would become the chairman of a regency council (but not regent). In this way, Sihanouk would retain actual power while still retaining the symbolic importance of the monarchy—of which he was the primary figure. In time, the regency council would fade from view.[10]

Second, Sihanouk did attempt to keep war from penetrating Cambodia. To do so, however, required Sihanouk to maneuver within the difficult waters of the Cold War. It is important to understand as Clymer argues, that Sihanouk's approach to the geographies of the Cold War was considerably different from that of the other major participants, such as the United States, North Vietnam, China, or the Soviet Union. By 1963 Sihanouk concluded that the US would not win in South Vietnam and therefore would not be the preeminent power in Southeast Asia. Thus in 1963 Sihanouk signed a treaty of friendship with the PRC; severed diplomatic relations with South Vietnam, and terminated all US assistance programs in Cambodia.[11] In addition, throughout the 1960s Sihanouk moved to nationalize Cambodia's banking and foreign trade and embarked on a program of setting up state-owned industries (a policy originally proposed by Hou Youn). Foreign investment, for example, was restricted and prices were fixed in an effort to ward off competition from imported goods.[12] As Chandler suggests, Sihanouk most likely expected France and China to pick up where the Americans left off in terms of aid provision. Sihanouk also perceived the necessity of arranging stronger diplomatic ties with other, potentially more important, states, namely the DRV and the PRC. Sihanouk therefore proceeded on the assumption (or perhaps, hope) that Hanoi would be able to control the Khmer Rouge, while concurrently the PRC would likewise be able to control the Vietnamese communists.

---

10 John Tully, *A Short History of Cambodia: From Empire to Survival* (Crows Nest, Australia: Allen & Unwin, 2005), 136–137.

11 Gareth Porter, "Vietnamese Communist Policy Toward Kampuchea, 1930–1975," in *Revolution and Its Aftermath in Kampuchea: Eight Essays*, David Chandler and Ben Kiernan (eds) (New Haven, CT: Yale University Southeast Asia Studies, 1983), 57–98; at 75.

12 Donald M. Seekins, "Historical Setting," in *Cambodia: A Country Study*, R.R. Ross (ed.) (Washington, DC: US Government Printing Office, 1990), 3–71; at 20.

Politically, such moves were insufficient and Sihanouk failed to obtain any support from the growing Leftist factions of the country. Moreover, Sihanouk's decision to nationalize Cambodia's banks and its import-export trade angered many of the business elite, including the Sino-Khmer commercial elite as well as the pro-western elite. Ironically, the nationalization of foreign trade encouraged the commercial elite of Phnom Penh to trade clandestinely with the Communist insurgents in Vietnam. Indeed, by 1967, over a quarter of Cambodia's rice harvest was being smuggled to these forces, who paid higher prices than the Cambodian government could afford.[13] Furthermore, Sihanouk's distancing of himself from the United States also frustrated Cambodia's military. This had the unfortunate effect of augmenting the militarization of Cambodian politics, signaled by the ascension of Sihanouk's defense chief, General Lon Nol, to prime minister in 1966.

In short, throughout the 1960s Sihanouk attempted to retain absolutist control of his country, while keeping the escalating Second Indochina War at bay. His neutralist stance, and contradictory policy initiatives, however, actually brought the political Right and Left together. As the decade wore on, the beleaguered monarch found himself confronted by an anti-royalist Rightist faction as well as an anti-royalist Leftist faction. And both factions began to make considerable inroads into Cambodian politics.[14]

Movements Transformed

In Vietnam, the scheduled elections of 1956 failed to materialize. Ngo Dinh Diem, the President of the newly-created State of Vietnam (South Vietnam), was an ardent nationalist and staunch anti-communist. However, he also saw little need for democratic principles and followed what he perceived as a mandate from heaven to rule the southern portion of Vietnam. In this belief, Diem's attitudes resonated well with the ambitions of American officials, namely President Dwight D. Eisenhower and Secretary of State John Foster Dulles. From the American's point of view, it was imperative to keep Vietnam permanently divided and to help Diem establish a noncommunist regime in Saigon.[15]

Following the conclusion of the Geneva Conference in 1954, Diem set out, with US assistance (meaning, money and advisors), to build South Vietnam. According to Patrick Heardon, three considerations were paramount in the minds of US advisors. First, there was the belief that if the South Vietnamese people enjoyed prosperity under the new regime, they would be less likely to support any

---

13 David P. Chandler, *A History of Cambodia*, 3rd edition (Boulder, CO: Westview Press, 2000), 201.

14 Porter, "Vietnamese Communist Policy," 77. To be fair, by the late 1960s Sihanouk had kept Cambodia out of the war and all but a few Cambodians from being killed. See Chandler, *Tragedy of Cambodian History*, 157. For an in-depth examination of Sihanouk's relations with the United States, see Kenton Clymer, *Troubled Relations: The United States and Cambodia since 1870* (Dekalb: Northern Illinois University Press, 2007).

15 Patrick J. Hearden, *The Tragedy of Vietnam: Causes and Consequences*, 2nd edition (New York: Pearson Longman, 2006), 54–55.

Communist movement. Consequently, many US advisors initially supported Diem with considerable enthusiasm. Second, if South Vietnam received financial support from the US, they would be able to obtain dollars that could be used to purchase Japanese goods. In this way, American actions would bolster both the geopolitical and geo-economic situations in Northeast *and* Southeast Asia. Third, policy-makers hoped that a successful South Vietnam could be used as a showcase for capitalism, thereby inducing those Vietnamese living in the north to abandon their support of Communism. Heardon concludes that while the minimum American goal was to erect a stable capitalist regime in South Vietnam, the maximum objective was to spark a counter-revolution in the DRV.[16]

Diem was a ruthless dictator. With US financial support, as well as the backing of the landlords and merchants, Diem moved quickly to suppress any political opposition that might challenge his rule. He moved to crush the southern Viet Minh who remained in the south, as well as the many religious sects that challenged his rule. A year after Geneva, Diem called for a national referendum to determine whether South Vietnam would remain a monarchy with Bao Dai as emperor, or become a republic with himself as president. Elections were held in October 1955, with Diem's own police "supervising" the balloting. Not surprisingly, Diem claimed that over 98 percent of the citizenry voted to support his ascension as president.[17]

Until July 1956, the Viet Minh who had stayed in the south held out hope that there would be a peaceful reunification of Vietnam following national elections. When these elections failed to materialize, however, the Viet Minh leaders decided that armed struggle was the only option to achieve national reunification. Ho Chi Minh and other Vietnamese Communist officials in the DRV, however, were wary of initiating armed revolution in the south. Their concerns were many and varied. They feared, for example, that overt war in the south would disrupt the economic rebuilding of the DRV. They worried also about provoking the United States into war if northern Communists were seen intervening in the south. In effect, the northern Communist leaders were asking their southern comrades to work toward political change through peaceful, democratic means, just as they were asking the Khmer communists to work more closely with Sihanouk.[18]

Continued, and harsh repression, however, impelled the Vietnamese Workers Party (VWP) to acquiesce to the demands of the south. By 1959 many Party workers, sympathizers, and former Viet Minh were being imprisoned, tortured, and assassinated by Diem's army and his American-trained police. Moreover, within this climate of intense repression, many high-ranking Vietnamese communists found

16 Hearden, *Tragedy of Vietnam*, 56.

17 Hearden, *Tragedy of Vietnam*, 59. According to Hearden, fraud and corruption were widespread; in some districts, more votes were recorded for Diem than there were registered voters. American officials, prior to the election, allegedly counseled Diem to "win" with a more credible 60 percent majority.

18 Hearden, *Tragedy of Vietnam*, 63.

temporary sanctuary in Cambodia.[19] Consequently, in 1960 the National Liberation Front (NLF) was established.[20]

The actions initiated by the Vietnamese Communists explains why the Khmer movement was encouraged to *not* engage in armed struggle against Sihanouk. Strategically, eastern and northeast Cambodia were identified as future "base areas" through which Vietnamese forces and supplied could be tunneled into South Vietnam—part of the "Ho Chi Minh trail." Initially, Vietnamese Communist bases in Cambodia were relatively small, consisting of storage and training facilities. Geographically, these were established just inside Cambodian territory. By 1966, these bases would expand dramatically to accommodate larger numbers of troops for longer periods of time. Spatially, in part because of South Vietnamese and American military offensives, these bases were pushed deeper into Cambodian territory. The continued use of the Ho Chi Minh trail, and the use of base areas, however, were reliant upon Sihanouk's support, or at least non-opposition to a Vietnamese presence on Cambodian territory. The KPRP was, from a Vietnamese point of view, to play a supportive—hence, secondary—role in Vietnam's struggle.[21]

For the Paris-educated Khmer Communists, the decisions of the DRV did not sit well. Given the on-going suppression of the Communist movement in Cambodia, such a strategy would spell their death. Equally important, the demands asked of their Vietnamese counterparts were seen as signs of continued Vietnamese aggression and dominance. Consequently, in 1960 a series of secret meetings among the Khmer Rouge were held in rail-cars in Phnom Penh.[22]

Resultant from the meeting, the KPRP was renamed as the Workers' Party of Kampuchea (WPK).[23] Symbolically, this name signified that the Khmer Party was on an equal level with that of North Vietnam. Concurrently, a new Central Committee was established. Tou Samouth, the long-time revolutionary, was elected as Party-Secretary, with Nuon Chea as Deputy. Saloth Sar emerged as the third-highest ranking official, serving as "assistant" to Samouth. Also placed on the Committee was Son Ngoc Minh, who was still residing in Hanoi.

The newly-established Central Committee offers some crucial insights into the transformations of the Cambodian Communist movement. The composition of the committee still favored those with ties to the ICP, and Saloth Sar was the only member who had participated in the Paris Marxist Circle. However, the ascension

---

19 Chandler, *Tragedy of Cambodian History*, 113.

20 In neighboring Laos, the Pathet Lao began also to engage in open armed struggle, having formed their own party, the People's Party of Laos, in 1955.

21 Chandler, *Tragedy of Cambodian History*, 113.

22 The origins, details, and consequences of the 1960 meetings continue to generate much debate among scholars of Cambodia. Did Hanoi, for example, initiate the meeting? To what extent, moreover, did the Vietnamese influence or dictate, the agenda? See Ben Kiernan, *How Pol Pot Came to Power: A History of Communism in Kampuchea, 1930–1975* (London: Verso, 1985), 188–194; David P. Chandler, *Brother Number One: A Political Biography of Pol Pot*, revised edition (Chiang Mai, Thailand: Silkworm Press, 2000), 59–60.

23 See Chandler, *Brother Number One*, 59. Elsewhere, it is referred to as the Khmer Workers' Party; see Chandler, *Tragedy of Cambodian History*, 114. In 1966 this Party was renamed the Communist Party of Kampuchea, or CPK.

of Saloth Sar, in particular, signaled the beginning of the Paris-educated group's entry into the highest echelons of the Cambodian Communist movement. Over time, they would gradually eliminate (murder) the veteran leadership; by 1977 Saloth Sar, Ieng Sary, and Nuon Chea were the only survivors from 1960 still serving on the committee. The others had been purged.[24]

The Paris-educated group had relatively little experience or interactions with the Cambodian populace. Most of these revolutionaries had led privileged lives, certainly in comparison to the "ordinary" Cambodian peasant. Many, including Saloth Sar, originated from land owning families and/or had royal connections. Most of these young men had received, or at least had access to, the best educational opportunities available in Cambodia. Also significant is that none were founding members of the resistance movement. They had been largely separate from the anti-French colonial movement, a factor that tremendously influenced their attitudes toward the Cambodian monarchy. The Paris-educated revolutionaries were elitist in orientation and resented the authoritarianism of Sihanouk.[25]

Most important, however, was the different ideologies (and hence strategies) forwarded by the Paris-educated revolutionaries and the veteran leaders. According to Kiernan, the main difference was in their contrasting perceptions of Sihanouk and of the necessity of working with Sihanouk. By extension, the relationship with Sihanouk greatly informed relations with the Vietnamese Communist movement. Saloth Sar and his supporters tended to be implacably opposed to Sihanouk and the monarchy in general. Under Sihanouk, Cambodia was neither independent nor free. The monarchy, furthermore, impeded the modernization of Cambodia that the revolutionaries sought. They also clarified the need for their own revolution, one that was not subordinate to the Vietnamese struggle. The Khmer veterans, conversely, were more inclined to hold sympathetic attitudes toward Vietnamese and of the greater Indochina struggle against the United States. Consequently, the veterans perceived the need to respect and indeed support Sihanouk's policy of neutrality. The objectives were two-fold: first, to not drive Sihanouk into the American camp and therefore enlarge the growing conflict and pull Cambodia into the struggle against American forces and, second, to keep the Communist bases within Cambodia open.[26]

Within two years the Paris-educated Khmers further entrenched themselves within the Cambodian revolution. In 1962 Tou Samouth disappeared.[27] His apparent death resulted in the ascension of Saloth Sar as Acting Secretary-General of the Party. Months later, at a secret meeting held in Phnom Penh, Saloth Sar was elected as Secretary-General. Nuon Chea retained the No. 2 position on the Politburo, and

---

24 Chandler, *Tragedy of Cambodian History*, 114.

25 Ben Kiernan, "Origins of Khmer Communism," *Southeast Asian Affairs* (1981): 161–180; at 176–177.

26 Kiernan, "Origins of Khmer Communism," 177; see also Kiernan, *How Pol Pot Came to Power*, 188–194; Becker, *When the War was Over*, 93.

27 It remains unclear as to whom arranged for Tou Samouth's death. It is possible, for example that Sihanouk's security police assassinated Tou Samouth, possibly with the assistance of Sieu Heng. Alternatively, Saloth Sar himself may have been involved in eliminating his superior.

was joined by Ieng Sary, So Phim, and Vorn Vet. Significantly, of the twelve Central Committee members, former Paris educated students held the positions 1, 3, 5, 6, and 11. Son Ngoc Minh, still in absentia, remained on the Committee, while long-time veterans Keo Meas (previously No. 6) and Non Suon were dropped altogether. Other prominent veterans also failed to be included in the Central Committee.[28] The ascension of the Paris-education group and their supporters[29] signaled drastic changes in the geographic organization of the Khmer Communist movement. Whereas the veteran leaders were dispersed to the most remote regions of the country, the Saloth Sar clique (including Ieng Sary, Nuon Chea, and Vorn Vet) assumed responsibilities for national activities.[30]

Student Protests and Revolutionary Shifts

In *The Wretched of the Earth*, Frantz Fanon observed that the "more the people understand, the more watchful they become, and the more they come to realize that finally everything depends on them and their salvation lies in their own cohesion, in the true understanding of their interests, and in knowing who their enemies are."[31] Fanon's point, of course, speaks to the understanding that class and social relations, economic systems, political structures, and so forth, are not natural but rather a function of particular institutions. Consequently, education and educational systems figure prominently in the generation of discontent.

During the 1950s and 1960s educational systems enlarged greatly throughout Cambodia. By 1969, for example, more than eleven thousand Khmers were attending Cambodian universities, and at least two million were enrolled in primary and secondary schools.[32] According to Tully, it had "produced the largest numbers of secondary and tertiary educated Khmers in the country's history, and many of them considered themselves intellectuals."[33] These students, however, increasingly voiced their dissatisfaction with the current state of affairs. Chandler explains that problems arose "because as segments of Cambodian society woke up, Sihanouk continued to use the machinery of absolutist rule to keep himself in power. His so-called children became more alert and better informed, but he allowed them no more freedom."[34]

---

28 Kiernan, *How Pol Pot Came to Power*, 200–201. So Phim apparently challenged Saloth Sar for the post of secretary-general. He was later purged.

29 Nuon Chea was in fact a Party veteran and did not receive any education in France. He received his education in Thailand and through his Thai connections had joined the Communist Party of Thailand (CPT). He later joined the ICP. Unlike other veterans, however, Nuon Chea was prepared to work with Pol Pot, perhaps signaled by his agreement to step down as Deputy Secretary in favor of Pol Pot in 1961, or by declining to run against him in 1963. Another veteran that garnered favor with the Paris-educated group was Mok, a former teacher who had developed a close relationship with Pol Pot. See Kiernan, *How Pol Pot Came to Power*, 201.

30 Kiernan, *How Pol Pot Came to Power*, 202.

31 Frantz Fanon, *The Wretched of the Earth* (New York: The Grove Press, 1963), 191.

32 Chandler, *Tragedy of Cambodian History*, 123.

33 Tully, *Short History of Cambodia*, 141.

34 Chandler, *Tragedy of Cambodian History*, 123.

Tully concurs, adding that for most of these educated Khmers, there were few jobs for this growing pool of comparatively well-educated youth. Students, in particular, began to sense that the government had failed them. They desired justice and social change.

An early indication of the growing unease among Cambodia's youth occurred in 1963. In Siem Reap tensions had grown for months between high school students and local authorities. Violence though soon erupted in February of that year. The proximate issue, innocently enough, was the right of school children to cycle along footpaths in the town. However, during the ensuing demonstrations it was discovered that a student had died in police custody. Protestors charged police brutality and corruption and demanded an investigation into the student's death. The authorities, however, refused to investigate. Over a thousand students stormed the police headquarters. Several students and police were killed. Over the following weeks, solidarity demonstrations appeared in Phnom Penh and Battambang.[35]

Accusations as to who instigated the unrest were plentiful. Some Khmer officials pointed the finger at the marginal Son Ngoc Thanh. Lon Nol, for example, provided Sihanouk with dossiers implicating "certain elements" among the students who were supposedly allied with Thanh. The United States, not surprisingly, accused Communist infiltrators and subversives. Sihanouk, for his part, was reluctant to blame *foreign* communist activity. In fact, Sihanouk had been in Beijing at the time of the riots, singing the praises of the PRC. On his return on March 1, Sihanouk berated the Prime Minister for incompetence, dismissed the government, announced the dissolution of the Sangkum and of parliament, and ordered new elections. Troops were likewise dispatched in an attempt to restore order.[36]

On March 4 Sihanouk published a list of thirty-four known and suspected Leftists. The names had been compiled by Lon Nol and his security police, based on their ongoing surveillance of journalists and teachers in the private schools. Included were Saloth Sar, Ieng Sary, and Son Sen. Absent, however, were other party members such as Ta Mok, Vorn Vet, and Nuon Chea.[37] Sihanouk charged these individuals as "cowards, hypocrites, saboteurs, subversise agents and traitors" and challenged them to form a cabinet and to govern the country. Three days later he upped the ante, summoning the entire group to a meeting at the Prime Minister's Office, where each man was to put in writing whether or not he agreed to Sihanouk's demands.[38] It would have been foolhardy to oppose Sihanouk openly at the time. Police guards had been posed outside the homes of those listed, and the schools where they taught were placed under surveillance.[39] Fearing their own assassination,

---

35 Short, *Pol Pot*, 142.

36 Short, *Pol Pot*, 142; Chandler, *Tragedy of Cambodian History*, 125.

37 Chandler, *Tragedy of Cambodian History*, 128. The names of Saloth Sar and Ieng Sary were probably included because they taught at Leftist schools. At the time, the actual "identity" of Saloth Sar was unknown to Sihanouk. Other notable persons on the list included Keng Vannsak, Hou Yuon, Khieu Samphan, and Hu Nim.

38 Quoted in Short, *Pol Pot*, 142–413.

39 Short, *Pol Pot*, 143.

32[40] of the men pledged their support for Sihanouk as the only man capable of moving the country.[41]

Following Sihanouk's public performance, events settled down. None of the 34 Leftists was arrested, and most continued to go about their day-to-day activities as before. However, for individuals such as Saloth Sar, the events served as a warning that it was time to shift the base of the revolution away from Phnom Penh.

Saloth Sar and other Party leaders had, in fact, been contemplating a withdrawal to the countryside for many months. It was apparent, for example, that political and social change would not be forthcoming through open and democratic processes. Although Sihanouk was being squeezed from both ends of the political spectrum, he was still a powerful (and brutal) opponent. On March 31 Saloth Sar left the city, followed by Ieng Sary two weeks later. They initially arrived at Snuol, a commune on the border of Kratie and Kompong Cham. From there they proceeded to a South Vietnamese Communist encampment along the Cambodian-Vietnamese border. Over time, as Sihanouk continued its ongoing, though at times muted, suppression of Khmer Communists, Saloth Sar and Ieng Sary were joined by other Party members. For the next seven years the Khmer Communists remained in hiding, moving from makeshift camps in eastern and northeastern Cambodia. There, they began preparing for a peasant revolution and; to fight guerrilla war against the Prince who still enjoyed widespread rural support.[42]

Deep within the mountains of northeastern Cambodia, Saloth Sar and the other Party leaders were isolated from events transpiring in Phnom Penh and the rest of the world. However, two events would significantly affect the political development of Saloth Sar and hence, the trajectory of the Cambodian Communist revolution. In 1965 Saloth Sar, Keo Meas, and several other Cambodian Communists, were summoned to the DRV for consultations. According to Chandler, the occasion of the meeting was most likely to discuss the role of the Cambodian Communist movement within the context of the escalating war in Vietnam. From the Vietnamese perspective, the top priority for all of Indochina was to defeat the United States. At the outset, the Khmer Communists were berated for pursuing a nationalist agenda, and were informed by Le Duan, secretary of the Vietnamese Communist Party, to subordinate Cambodia's interests to those of Vietnam's. First, the Khmer Communists were asked to work with Sihanouk. Throughout the previous two years, Sihanouk had been moving closer to both the DRV and the NLF. Indeed, as indicated earlier, Sihanouk had rejected all US military assistance. Frustratingly for Saloth Sar, however, was that Sihanouk had also continued his brutal suppression of the Khmer Communists. However, from the perspective of the North Vietnamese, continued support for Sihanouk was to take precedence. Sihanouk had reached an agreement with the DRV that allowed the NLF and other insurgents to station themselves on Cambodian territory. In exchange, the

---

40 Saloth Sar and Chou Chet did not appear; Saloth Sar went into hiding when the list was published and Chou Chet, who had been released from prison three weeks earlier, also fled. See Short, *Pol Pot*, 143.

41 Short, *Pot Pot*, 143; see also Chandler, *Tragedy of Cambodian History*, 126–127.

42 Kiernan, "Origins of Khmer Communism," 178.

DRV pledged to honor Cambodia's independence and borders when the war was over. Thus, the DRV looked to the Khmer Communists to minimize their attacks against Sihanouk. Second, the Khmer Communists were asked to play a supporting role. While postponing armed struggle in Cambodia, the Khmer were expected to provide military and logistical support for the Vietnamese.[43]

Following his visit to Vietnam, Saloth Sar next traveled to Beijing and North Korea. There, his reception was considerably different. Sar's arrival in China was pivotal. His visits were coincident with the early phases of Mao Zedong's Cultural Revolution, a momentous event that would considerably influence Saloth Sar's promotion of Democratic Kampuchea as a Communist Utopia. According to Chandler, Sar was drawn to Mao's emphasis on "continuous class warfare, individual revolutionary will, and the importance of poor peasants—ideological areas in which China differed from Vietnam and which were later emphasized in Democratic Kampuchea." Furthermore, Chandler contends that Sar must have been impressed by the scale, autonomy, and momentum of China's social mobilization as he had been by Yugoslavia's in 1950. It was, as Chandler concludes, China's triumphant revolution rather than Vietnam's arduous, unfinished struggle that Sar wished to bring back to Cambodia.[44] More immediately, however, the visit was important because of the course of direction the Chinese encouraged their Khmer counterparts to follow. Specifically, Saloth Sar was probably told to work with Sihanouk and to support the Vietnamese. This could only have been construed by Saloth Sar as a slight. More than likely, Saloth Sar and the other Khmer Communists probably felt sacrificed by their Communist brethren. Compounding these decisions, the CPK received minimal assistance from the North Vietnamese, the Chinese, or the Soviets. Simply put, DRV leaders were reluctant to alienate the embattled Sihanouk since he permitted vital supplies to pass through Cambodian territory. China and the USSR likewise were providing Sihanouk with arms—many of which, ironically, were used against the communist insurgency.[45]

The Samlaut Rebellion

Located in the province of Battambang, the village of Samlaut lies ten miles from the Thai border, and 15 miles southwest of Battambang City. Historically, Battambang province was Cambodia's rice-bowl; the region had also provided refuge for the Viet Minh and Issarak guerrillas during the first Indochina War. It was here that Sihanouk's policies ignited a peasant revolt and changed the direction of the Communist Revolution.

After independence Sihanouk chose Battambang as an area where landless peasants from the southwest, as well as Cambodian refugees from South Vietnam, could be resettled. The newcomers, many of whom arrived in 1965–1966, received small but not insignificant government subsidies. Merchants and officials soon

---

43 Chandler, *Brother Number One*, 69–70; Chandler, *Tragedy of Cambodian History*, 147–148.

44 Chandler, *Brother Number One*, 72–73.

45 Seekins, "Historical Setting," 42.

bought up large portions of land, and many long-time residents found themselves displaced and in debt. Local peasants, furthermore, were angered by the corruption of local officials. They resented the authorities' demands for free labor and "voluntary financial contributions" to carry out government projects.[46] And in 1966–1967 the government altered its rice collection system.

As indicated earlier, Sihanouk since 1963 had refused American aid, nationalized the import-export sector of the economy, and closed Cambodia's privately owned banks. His decisions were based on the hope that by minimizing ties with the United States and fostering good relations with the Communist bloc, he might prevent the expansion of the war into Cambodia. However, his cutting of ties to the US resulted in a loss of substantial funds that had balanced his budget for years. No other donor (i.e., the PRC or the USSR) was able to match the lost funds and Sihanouk reduced the military budget to cover his losses, thereby ensuring the ire of the military. Furthermore, the nationalization of foreign trade encouraged the commercial elite to trade clandestinely with Communist insurgents in Vietnam. The escalation of the war in Vietnam translated into increased troops levels. Recruitment into the NLF in southern Vietnam, for example, quadrupled in one year, rising from 45,000 new recruits in 1964 to 160,000 in 1965. These troops needed to be fed. The off-shoot was that large amounts of Cambodian-grown rice were smuggled into Vietnam to feed the growing armies; indeed an estimated two-thirds of the country's rice exports were being smuggled out of the country through illegal sales.[47] Cambodia's economy, which was dependent on the taxes of rice exports for its revenue, plunged toward bankruptcy.

In 1967 a decision was made that army units would gather the rice surplus in several areas, pay government prices for it, and transport it to government warehouses. Through such a measure, merchants would then be able to sell the rice abroad at a higher price, thereby increasing profits for the national treasury. Such a move, however, generated intense peasant anger. At the time, the Vietnamese were paying double the official rate. Many farmers began to deliberately withhold their harvests, while others refused to grow more than was needed for their own households.[48] In response, the government initiated a new collection system, called *ramassage du paddy*. Under this system, peasants who withheld grain would be forced at gunpoint to sell the rice to government at lowered official prices. To ensure the practice, Lon Nol replaced provincial officials (who had been appointed by Sihanouk) with his own men.[49]

By February 1967 tensions had reached a breaking point. A peasant uprising occurred in the town of Pailin on the border with Thailand and anti-government demonstrations erupted in Battambang City. Sihanouk blamed the Khmer Communists for the disturbances, accusing Khieu Samphan, Hou Youn, and Hu Nim of organizing the peasant revolts. Khieu Samphan had, indeed, seized the issue to organize a demonstration to pressure the government to end the *ramassage*.

---

46 Short, *Pol Pot*, 165; Chandler, *Tragedy of Cambodia*, 164.

47 David Chandler, *A History of Cambodia*, 3rd edition (Boulder, CO: Westview Press, 2000), 201; Becker, *When the War was Over*, 103.

48 Becker, *When the War was Over*, 101.

49 Becker, *When the War was Over*, 103.

However, the uprisings (similar to those in Siem Reap four years earlier) appear to have been spontaneous.[50] Sihanouk, for his part, could not believe that his loyal subjects, his "children," would move against him because of genuine grievances connected with his policies.[51]

On April 2, two soldiers collecting rice were murdered at Samlaut and a number of rifles captured. This event precipitated a larger protest movement. Soon, a crowd of approximately two hundred began to coalesce. The protestors than marched on a youth agricultural settlement near village of Kranhoung, burning it to the ground. By nightfall two other army posts had been attacked and a local official killed. Sporadic fighting continued for the next four days.

Sihanouk responded in force. In an effort to punish the rebels, Sihanouk mobilized the army and air force to launch a counter-attack. In the ensuing melee, tens of thousands of peasants fled the region, pursued and killed by government forces. Sihanouk offered a bounty for the severed head of any Cambodian Communist or rebel that was captured. Even larger sums were offered for each rebel village burned to the ground. Suspected instigators were publically executed and films of the executions were shown throughout the country. In yet another tragic irony that befell Cambodia, military planes that had been provided by Communist China were used to strafe and bomb suspected rebel and Communist villages across the Cambodian countryside. Years later, Sihanouk acknowledged—unapologetically—that as many as 10,000 peasants were killed in retribution.[52]

The uprising at Samlaut signified both the weakening of Sihanouk's grip on the country as well as the limitations and contradictions of his policies. Believing, for example, that the rebellions were instigated by Khmer Communists, Sihanouk was angered by what he interpreted as a betrayal by the DRV. Throughout the 1960s Sihanouk had held out hope that the Vietnamese Communists would be able to restrain their Cambodian counterparts and that an alliance with the North would prevent war from infringing on Cambodian territory. Consequently, Sihanouk had agreed to Hanoi's demands that would permit the Vietnamese Communists rights of passage to Sihanoukville, the deep-sea port built on the Gulf of Siam to prevent the US from blockading his country. Vietnamese Communists were allowed to ship arms to Sihanoukville and transport them overland to eastern Cambodia.[53] Following the uprising at Samlaut, however, Sihanouk sensed that the DRV was in no position to control the actions of the Khmer Rouge.

Ironically, the DRV had not betrayed Sihanouk and were not responsible for the uprising. In fact, the DRV continued to oppose any Communist armed struggle in Cambodia, for fear of widening the war in Vietnam. And in a double irony,

---

50 For an alternative perspective, see Kiernan, *How Pol Pot Came to Power*.

51 Chandler, *History of Cambodia*, 201.

52 Becker, *When the War was Over*, 105; Short, *Pol Pot*, 167; Chandler, *Tragedy of Cambodia*, 165; Tully, *Short History of Cambodia*, 147.

53 Ironically, Lon Nol—Sihanouk's military chief—was in charge of overseeing the Communist transport of arms from Sihanoukville to southern Vietnam. In the process, Lon Nol skimmed off amounts of the arms, reportedly with Vietnamese approval, as a price for cooperation. Lon Nol would then use these weapons in his fight against the Cambodian Communists and, later, against ethnic Vietnamese. See Becker, *When the War was Over*, 102.

the Khmer Rouge likewise were not responsible for the rebellion. As the events unfolded, Saloth Sar and the Khmer Communists were caught unaware. Indeed, the Khmer Rouge suspected that the rebellion was initiated by those on the political Right who also opposed Sihanouk. Most likely, the unrest was spontaneous, an unplanned social movement fostered by a combination of rising expectations coupled with a decaying economy and increasing governmental corruption. Chandler explains that the rebellion at Samlaut was the product of "local grievances against injustice and social change, corruption, and ham-fisted government behavior." He does acknowledge, however, that Leftist teachers and students from Battambang City undoubtedly encouraged people to blame their troubles on feudalism, Lon Nol, and the United States but these criticisms "did not respond to orders from the CPK central committee to attack Sihanouk's government directly."[54]

The various perceptions of culpability contributed to the far-reaching consequences of the uprising. For the Khmer Rouge, the Samlaut rebellion proved that armed struggle was not yet viable. From a military perspective, the brutality which Sihanouk used to suppress the peasants was not lost. However, from a political standpoint, lessons were also learned. The anger evinced by the peasants, combined with Sihanouk's repression, provided both a model and precedent and thus made it easier to recruit new members to the Khmer Rouge.[55] In addition, that the uprisings occurred in Battambang proved fortuitous for the Khmer Rouge in other ways. As Becker explains, this was probably the one province in the entire country where conditions for revolution were ripe. Here, for example, the peasants were at the mercy of large landholders, they had experienced considerable displacement, and the local government was notoriously corrupt.[56]

Sihanouk's perception of the uprising likewise set in motion a series of events that had unanticipated consequences. Publically, Sihanouk accused Khieu Samphan, Hou Youn and Hu Nim of masterminding the rebellion. Fearing for their lives in an expected anti-Communist pogrom, Hou Youn and Khieu Samphan fled the capital in late April, followed later in the year by Hu Nim. Many observers, both in Cambodia and elsewhere, assumed that the three Communists had been assassinated by Lon Nol's forces—a rumor that Sihanouk, the United States, and the Khmer Rouge all used to their own advantages. For Sihanouk and the United States, this was an indication that the rebellion and those who instigated it had been defeated. For the Khmer Rouge, Hou Youn, Hu Nim, and Khieu Samphan were held as martyrs, men who had sacrificed their lives in the quest for social justice. Student opinion, consequently, hardened against Sihanouk.[57] Later, when Hu Nim, Hou Youn, and Khieu Samphan reappeared, they became known as the "Three Ghosts."

---

54 Chandler, *Tragedy of Cambodian History*, 166.
55 Chandler, *Tragedy of Cambodian History*, 167.
56 Becker, *When the War was Over*, 103.
57 Chandler, *Brother Number One*, 78; Tully, *Short History of Cambodia*, 147.

The Encroaching War

Look, we're not interested in Cambodia. We're only interested in it not being used as a base.

US National Security Advisor Henry Kissinger[58]

Despite Sihanouk's efforts—or, arguably, because of his efforts—Cambodia was increasingly drawn into the war. As the United States intensified its military actions in Vietnam, both the military forces of the DRV and the NLF intensified their use of Cambodian territory as sanctuary from the armed forces of the Republic of Vietnam and the United States. Communist troops, moreover, constructed an elaborate supply route—the Ho Chi Minh Trail—to funnel troops and supplies to the battlefields of southern Vietnam. As Vietnamese communist forces sought refuge in the eastern fringes of Cambodia, American and South Vietnamese forces followed.

In 1967 American military advisors initiated Operation Salem House. Whereas American advisors had been in Cambodia, clandestinely, since the early 1960s, Salem House systematized these operations. Teams of six to eight Americans and South Vietnamese would enter Cambodia seeking tactical intelligence. At first, these teams were severely limited in terms of their geographic coverage, restricted to the northeastern tip of Cambodia. Over time, however, these operations (renamed Daniel Boone) expanded to encompass the entire Cambodian-Vietnamese border region. By October 1968 the number of covert missions had increased both in scale and scope. The limitations on the number of Americans who could be included were removed, and the use of anti-personnel mines—which inflicted a devastating toll on the Cambodian people—was permitted. By 1969 US forces had conducted 454 covert missions in Cambodia. During the following year, the number of missions increased to 558. By the time the missions were ended in 1972, over 1,885 operations had been conducted.[59] Estimates of Cambodian civilians killed between 1969 and 1973 run as high as 150,000.[60] And the result of these operations, aside from the devastating toll on Cambodian life? Twenty-seven Americans were confirmed dead.[61] Only 24 alleged prisoners were captured.

Cambodian officials, including Sihanouk, protested these missions. In just one month—October 1969—representatives of Cambodia protested 83 separate incidents of American intervention. Aerial and artillery attacks, ostensibly targeting NLF strongholds, were more often destroying Cambodian villages—houses, schools, bridges—and killing more Cambodian civilians than any enemy personnel. Farming was disrupted and livestock was also killed.[62] Strategically, the expansion of attacks on Cambodian territory simply drove the NLF and NVA deeper in Cambodia.

---

58 Quoted in Shawcross, *Sideshow*, 145.

59 Clymer, *Troubled Relations*, 99–100.

60 Clymer, *Troubled Relations*, 137.

61 Relatives of the killed soldiers were never informed that their loved ones had died in Cambodia.

62 Clymer, *Troubled Relations*, 99. Sihanouk *did not* give his approval for the military encroachment on Cambodia's territory. To be sure, Sihanouk tolerated limited infringement on his country's territory; he did acquiesce to certain demands of both the DRV and the US

As the war in Vietnam ground on, US military officials requested, repeatedly, permission to carry the war—overtly—into Cambodia and Laos. And repeatedly, President Johnson refused such requests. President Nixon, however, felt less constrained to expand the war. Widely known for his hawkish views toward war, Nixon won the presidency with a promise to "end the war and win the peace."[63] In actuality, Nixon's objective—and one shared by his equally hawkish National Security Advisor, Henry Kissinger—was to end the war while retaining American credibility in Southeast Asia and beyond. And if Cambodia had to be sacrificed to reach these objectives, so be it. Cambodia was, as Shawcross describes, a "sideshow" to the American war efforts in Indochina. However, Cambodia was to play a more tragic role in this performance of geopolitical machismo.

As part of Nixon's overall approach to the war in Indochina, decisions were reached to eliminate suspected NLF sanctuaries within Cambodia. On February 9, 1969 General Creighton Abrams—who had replaced General William Westmoreland as commander of American forces in Vietnam—recommended a single, intensive attack on a site that was presumed to be the headquarters of NLF forces, identified as the Central Office for South Vietnam (COSVN). Previously, Johnson had refused such requests but the strategy resonated with the more hawkish Nixon.

On March 18, 1969 Nixon ordered a series of secret and illegal B-52 bombings to be conducted on Cambodian territory. For more than a year, from March 1969 through May 1970, over 3,800 B-52 sorties were flown over Cambodian territory, dropping more than one hundred thousand tons of bombs.[64] The death toll of civilian casualties is unknown, with estimates ranging from 30,000 to over 500,000.

Strategically, the bombing was ineffective and unwarranted. The aerial attacks did not seriously affect the ability of the NLF to continue its operations.[65] When news of the bombings became public, however, American officials claimed that the operation was a success, and that the targeted areas were not inhabited by Cambodian civilians. Indeed, Kissinger stated that the areas bombed were "unpopulated," a lie that Nixon often repeated.[66] Furthermore, both Nixon and Kissinger would later claim that Sihanouk approved of, or perhaps even encouraged, the air strikes. However, as William Shawcross explains, whether or not Sihanouk agreed is irrelevant. As Shawcross writes, the "whims of, and the constraints upon, a foreign prince are not the grounds for [an American] President to wage war."[67] Regardless of Sihanouk's position on the bombings, the actions of both Nixon and Kissinger reflect that Cambodia was simply a surrogate for other geopolitical goals. Nixon, for example,

---

to operate within Cambodia's borders. However, as Clymer explains, these were specific and limited arrangements, agreed to in a desperate hope to keep the violence away from Cambodia and to retain his country's independence and neutrality.

63 Nixon had advocated American intervention in Vietnam as early as 1954; he also refused to rule out the possibility of nuclear weapons in war against Vietnam or the Chinese.

64 Clymer, *Troubled Relations*, 96.

65 Clymer, *Troubled Relations*, 96–101.

66 Clymer, *Troubled Relations*, 95.

67 Shawcross, *Sideshow*, 94.

claimed in his memoirs that the second Menu attack (Lunch) was retaliatory for North Korea's downing of an American EC-121 spy plane.[68]

Nixon's personal escalation of the war was given impetus by the 1970 coup against Sihanouk. While traveling to France, Sihanouk had entrusted his government to Lon Nol and his pro-western deputy prime minister, Prince Sisowath Sirik Matak. According to Chandler, Matak—Sihanouk's cousin—was the most prominent of the plotters. Matak was impatient with Sihanouk's mismanagement of the economy and was dismayed by the presence of Vietnamese bases on Cambodian soil.[69] While Sihanouk was away, Lon Nol and Matak launched attacks on the Vietnamese communist positions, organized anti-Vietnamese demonstrations, and reestablished ties with various non-communist groups. Sihanouk, upon hearing these actions, condemned the moves of Lon Nol and Matak. In response, Matak pressured Lon Nol to lead a coup against Sihanouk and, on March 18, 1970, the National Assembly voted 89–3 to depose the ruler.[70]

With Sihanouk removed, and the more pliable and pro-American Lon Nol in power, Nixon hurriedly expanded the American presence in Cambodia. On April 30, 1970 Nixon announced to the American public that US ground forces, accompanied by the Army of the Republic of Vietnam (ARVN) had made a strategic "incursion" into Cambodia. Nixon explained that "North Vietnam [had] increased its military aggression," especially in Cambodia, and that "to protect [Americans] who are in Vietnam and to guarantee the continued success" of US operations, Nixon concluded that the time had come for action. Nixon then proceeded to misinform the American public that "American policy [had] been to scrupulously respect the neutrality of the Cambodian people." He failed to mention, however, the covert operations dating back to 1967, or the bombings associated with Operation Menu. Rather, Nixon told his listeners that the United States had "maintained a skeleton diplomatic mission of fewer than 14 in Cambodia's capital" and that "for the past 5 years [the United States had] provided no military assistance whatever and no economic assistance to Cambodia." North Vietnam, Nixon asserted, was guilty of interference in the sovereignty of Cambodia. North Vietnam had established military sanctuaries along the Cambodian border with South Vietnam, and these areas contained "major base camps, training sites, logistics facilities, weapons and ammunition factories, airstrips, and prisoner-of-war compounds." Nixon explained, further, that Cambodia "sent out a call to the United States" for assistance. And without American support,

---

68 Shawcross, *Sideshow*, 92.

69 Chandler, *History of Cambodia*, 208.

70 Shawcross, *Sideshow*, 122. Although there is no concrete evidence that the United States orchestrated the removal of Sihanouk, it is known that American officials had always found the Sihanouk regime inadequate and troublesome. Furthermore, the Pentagon had long identified the political and right-leaning elite of Cambodia, as well as the officer corps, as the groups that would best serve US interests. These two groups, represented by the coup leaders Matak and Lon Nol, thus had no reason to doubt that their actions would be acceptable to Washington. Clymer concludes, therefore, that there was likely some American foreknowledge of the coup, and that if the CIA was not involved, American military intelligence officials in Vietnam probably were; See Clymer, *Troubled Relations*, 102.

Cambodia would become a vast enemy staging area and a springboard for attacks on South Vietnam.[71]

During the course of Nixon's speech, he stressed that this was not an invasion of Cambodia. He explained that the purpose of the "incursion" was not to occupy the areas but merely to drive the Vietnamese communists out of Cambodia. And wary of Soviet or Chinese responses, Nixon announced that the actions were "in no way directed to the security interests of any nation" and that any government that chose to use the incursion as "a pretext for harming relations with the United States" would be doing so on its own responsibility, and on its own initiative." Nixon then reaffirmed that the United States undertook the incursion "not for the purpose of expanding the war into Cambodia, but for the purpose of ending the war in Vietnam and winning the peace."[72] Orwellian in his thinking, Nixon maintained that peace could only be achieved through an escalation of war. As Clymer concludes, the invasion brought Cambodia itself directly into the war—and with devastating consequences.[73]

The invasion, code-named Operation Shoemaker, involved more than 44,000 ARVN and American troops, and was concentrated along the Cambodian-Vietnamese border. When the dust had settled, and the bodies counted, Nixon claimed publically that over 11,000 Vietnamese Communists had been killed. Internal reports suggest that these estimates were exaggerated. Significantly, what Nixon failed to mention was that the "incursion" played into the hands of the Cambodian Communists.

Following the coup, Sihanouk had been persuaded by the Chinese Prime Minister Zhou Enlai to form a military alliance with the Vietnamese and Cambodian Communists and to lead a government-in-exile.[74] Sihanouk, as Arnold Isaacs explains, was far too clear-eyed not to have realized, even in these early weeks, that he was tying himself to interests that were mortally dangerous to Cambodia's survival.[75] However, without receiving support from the United States, Sihanouk

---

71 Richard M. Nixon, "Address to the Nation on the Situation in Southeast Asia," April 30, 1970, www.nixonlibrary.org.

72 Nixon, "Address to the Nation."

73 Clymer, *Troubled Relations*, 109.

74 Initially, China attempted to align itself with the Lon Nol government if three conditions were met: permission for the Chinese to continue supplying the Vietnamese Communists through Cambodian territory; authorization of Vietnamese Communists to maintain their bases inside Cambodia; and Khmer support of the Vietnamese Communists in government statements. In effect, the Chinese were willing to postpone the Cambodian revolution in order to help the Vietnamese revolution and to maintain a Chinese-Vietnamese front against the United States. They would, in Becker's words "sell out the Cambodian communists to ensure a defeat of the Americans in Vietnam." Arnold Isaacs explains further that Hanoi most likely would have preferred to work directly with the Lon Nol government. The most desirable outcome of the coup, from the perspective of the DRV, would have been to renew their lease on the sanctuaries, recreating that same accommodation they had enjoyed with Sihanouk. The Lon Nol government, given its anti-Vietnamese and anti-communist hard-line stance, predictably refused the Chinese overture. See Becker, *When the War was Over*, 117. See also Isaacs, *Without Honor: Defeat in Cambodia* (Baltimore: Johns Hopkins University Press, 1983), 201.

75 Isaacs, *Without Honor*, 199.

had few options. On March 23, 1970 Sihanouk announced the formation of the National United Front of Cambodia (*Front Uni National du Kampuchea*, or FUNK), a political and military coalition of Royalists and the Khmer Rouge committed to destroying the Lon Nol regime.[76] Later, in May, the Royal Government of National Union of Kampuchea (*Gouvernement Royal d'Union Nationale du Kampuchea*, or GRUNK) was announced. Sihanouk assumed the post of GRUNK head of state, and appointed Penn Nouth as prime minister. Khieu Samphan was designated deputy prime minister, minister of defense, and commander in chief of the GRUNK armed forces (Saloth Sar in fact commanded in the military). Hu Nim served as minister of information and Hou Yuon assumed multiple responsibilities as minister of interior, communal reforms, and cooperatives.[77] Not surprisingly, these developments were supported by the DRV and the USSR; both governments viewed the pending civil war as "just."

Given these political developments, the dependence of the Lon Nol regime on the Americans for military equipment and political support played into the Communists' hands.[78] Whereas the coup was largely popular with Cambodia's urban population, as well as the merchants, the military, and the extreme Right, those in more rural areas were generally not supportive. In fact, following the coup, many pro-Sihanouk riots erupted throughout the countryside. Sihanouk issued an appeal to the Cambodian people whereupon royalist supporters would join with the Khmer Rouge in a unified effort to remove the Lon Nol government. Throughout the ensuing months, as ARNV and American forces fought their way across the Cambodian landscape, thousands of young men and women rallied to Sihanouk by joining the Khmer Rouge. The Khmer Rouge Army—officially, the Cambodian People's National Liberation Armed Forces, or CPNLAF—increased dramatically.[79] Whereas in 1970 the Khmer Rouge forces were described as "marginal," by 1972 these forces were estimated to have grown to more than 20,000. Indeed, some US officials presented figures

---

76 Both Sihanouk and the Khmer Rouge recognized the tenuous basis of the alliance. The Khmer Rouge continued to hold Sihanouk responsible for the war in Cambodia, but well understood his popularity. Saloth Sar, consequently, used Sihanouk for propaganda and recruitment purposes. Sihanouk likewise understood that his role in the alliance was little more than a titular figurehead. He hoped, however, that he might use the arrangement as a means to deposing of Lon Nol and to eventually return to power. Such a move required the support of the United States, however, which was not forthcoming. See chapter 7 in Clymer, *Troubled Relations*.

77 Seekins, "Historical Setting," 43–44. Khmer Rouge members of GRUNK claimed that it was not a government-in-exile because Khieu Samphan and the other officials remained in residence in Cambodia. Publically, neither Saloth Sar, Ieng Sary, nor Nuon Chea were identified as top leaders. The formation of FUNK and GRUNK also saw the return of the "Three Ghosts."

78 Chandler, *Brother Number One*, 90.

79 The CPNLAF was the renamed Revolutionary Army of Kampuchea (RAK), which had been founded by Pol Pot in 1968. It was also known as the People's Liberation Armed Forces of Kampuchea (PNLAFK); see Seekins, "Historical Setting," 318.

between 35,000 and 50,000, with some CIA estimates placing CPNLAF strength at over 150,000.[80]

Nixon's decision to escalate the air war over Cambodia also augmented the Communist revolution. On July 27, 1970 B-52 raids resumed. Truong Nhu Tang, the NLF Justice Minister, described the consequences: "Nothing the guerrillas had to endure compared with the stark terrorisation of the B-52 bombardments ... It was as if an enormous scythe had swept through the jungle, felling the giant teak and *go* trees like grass in its way, shredding them into billions of shattered splinters ... It was not just that things were destroyed; in some awesome way, they had ceased to exist ... There would simply be nothing there, in an unrecognisable landscape gouged by immense craters."[81]

Terrorized by the intense bombing campaign, thousands of Cambodian peasants flocked to Phnom Penh and other provincial towns. Others fled into neighboring Vietnam, or attempted to escape to Thailand. Those who remained invariably joined the Khmer Rouge. As Short concludes, carpet bombings "gave the Khmer Rouges a propaganda windfall which they exploited to the hilt—taking peasants for political education lessons among the bomb craters and shrapnel, explaining to them that Lon Nol had sold Cambodia to the Americans in order to stay in power and that the US, like Vietnam and Thailand, was bent on the country's annihilation so that, when the war was over, Cambodia would cease to exist."[82] Recruitment was further propelled when, on October 9, 1970, Lon Nol formally abolished the monarchy and redesignated Cambodia as the Khmer Republic. According to Seekins, the concept of a republic was not popular with most peasants, who had grown up with the idea that something was seriously awry in a Cambodia without a monarch.[83]

The Consolidation of the Khmer Rouge

Lon Nol's coup against Sihanouk, coupled with his support of increased US military operations within Cambodia, significantly transformed both the conduct and the constitution of the Communist revolution. On the one hand, events in Cambodia altered the strategy of the Vietnamese Communists. First, the DRV asserted that Cambodia was no longer neutral and, consequently, the Vietnamese Communists sought to play a major role in establishing local revolutionary political and military organs inside Cambodia.[84] Second, Vietnamese forces sought to protect the ethnic Vietnamese residents living in Cambodia. Within weeks of the coup, Lon Nol

---

80 Clymer, *Troubled Relations*, 119. Significantly, the coup initially threatened the long-term success of the revolution. If the DRV and the PRC had been able to broker a deal with Lon Nol's government, the Cambodian Communist revolutionary movement would in all likelihood have been sacrificed once again for the greater "Indochina," or more properly, the Vietnamese, revolution. As it was, the DRV was compelled to wage war against the Lon Nol government, thereby indirectly strengthening the Khmer Rouge.

81 Quoted in Short, *Pol Pot*, 215.

82 Short, *Pol Pot*, 218.

83 Seekins, "Historical Setting," 44.

84 Porter, "Vietnamese Communist Policy," 83.

ordered his security forces to erect detention camps and holding centers. An extreme nationalist and anti-Communist, Lon Nol served tirelessly in both Cambodian political and military affairs. In 1946, as part of Cambodia's first national elections, Lon Nol founded his own political party, the Khmer Renovation Party and in 1955 he merged his party with Sihanouk's Sangkum Party. During the First Indochina War, at the request of Sihanouk, Lon Nol fought against the Issaraks and the rebel Vietminh. Lon Nol, not unlike the Khmer Rouge, sought to purify the Khmer race, culture, and religion of all foreign pollutants. The coup provided Lon Nol with his opportunity. Approximately 30,000 Vietnamese residents—many of whom had lived for generations in Cambodia—were rounded up and detained as prisoners. Most of these were charged as being sympathizers with the Vietnamese communists, who had used Cambodia as sanctuary from the war in Vietnam.[85]

Then the killings began. On April 10, 1970 soldiers of Lon Nol's army forcibly removed the men, women, and children from Prasot, a small village near the Vietnamese border. Approximately 100 civilians, all of whome were of Vietnamese heritage, were detained at a farming cooperative and then slaughtered. Two weeks after the attack on Prasot a Catholic Vietnamese settlement on the Chrui Changwar isthmus in the vicinity of Phnom Penh was targeted. Government soldiers appeared at night and rounded up approximately 800 Vietnamese laborers. The captives' hands were tied behind their backs and they were shoved onto waiting boats. The soldiers then executed every Vietnamese captive and threw the bodies into the Bassac River. Days later, former Southeast Asian correspondent for the *New York Times* Henry Kamm was present at the scene. He writes, "I took the ferry five days after the ghastly flow began and saw a group of five bodies tied together by their feet with rattan cord, as well as four bodies, their hands tied, floating singly toward Vietnam, about thirty miles downstream."[86] The massacre at Chrui Changwar was repeated in the town of Takeo, located south of the capital near the Vietnamese border. Here, approximately 150 Vietnamese men were rounded up and sequestered in a school. Each evening, the captives' wives and children brought food. One night, however, while the families were eating, soldiers fired point-blank into the mass of people. Estimates place the number of dead at 100. Trucks came to haul away the corpses.[87]

Consequently, to utilize the Cambodian revolution to support the Vietnamese revolution, and to protect ethnic Vietnamese, DRV forces assumed the brunt of fighting against Lon Nol and, when the Republican forces were defeated, these areas would be turned over to local Khmer leaders. The Khmer Rouge, concurrently, established other "liberated zones," particularly in the south and the southeast, where they operated independently of the Vietnamese. Slowly at first, and increasing by 1973, the Khmer Communists began their task of reforming Cambodia's society. As the Khmer Rouge took over administrative control from the Vietnamese, they imposed, in Isaacs's words, an "Orwellian rigor" on their cadres. New recruits, for example,

---

85 Becker, *When the War was Over*, 125.

86 Henry Kamm, *Report from a Stricken Land* (New York: Arcade Publishing, 1998), 79; see also Becker, *When the War was Over*, 125.

87 Kamm, *Cambodia*, 80.

were required to assume new names to symbolize the surrender of all individual consciousness and were trained in a blind submissiveness.[88] The end result was that while the Vietnamese carried the burden of military operations during the early 1970s, the Khmer Rouge forces were able to concentrate on political recruitment, training, and socialization.[89] Their forces, moreover, incurred fewer casualties on the battlefield. As a result, the Cambodian Communist troops became larger, better organized, more self-confident.[90]

On the other hand, many Cambodian Communists living in exile since 1954 saw this as an opportunity to actively rejoin the resistance within their country. Soon after the coup, approximately 1,000 Cambodians began the long and dangerous trek south, along the Ho Chi Minh trail, to reunite with their Communist brethren. Most were unaware, however, of the significant changes that had developed within the Cambodian Communist movement. Saloth Sar and his colleagues, conversely, viewed the Hanoi-trained Khmer with suspicion, but also as an opportunity. To be sure, the returning Khmer could, from Saloth Sar's perspective, prove a potentially invaluable resource. While in Vietnam, many had been extensively schooled in preparation of Cambodia's communist revolution. They had engaged in political studies of Marxism-Leninism and understood the Marxist-Leninist concept of revolution. Furthermore, many had received military training. The CPK recruits, conversely, were poorly trained, badly equipped, and largely unresponsive to discipline. Saloth Sar correctly saw the Hanoi-trained Khmer as a means to rectifying the limitations of his forces.[91] And the fact that the Vietnamese Communists continued to shoulder the burden of military operations permitted Saloth Sar time to exploit the returning Cambodian Communists. However, while viewed as a resource, the CPK viewed the returnees as a potentially subversive group. The Hanoi-trained Khmer were viewed as "Vietnamese" disguised in "Cambodian" bodies, an internal vanguard that threatened to wrest control of the Cambodian revolution from the CPK.

These two perceptions of the Hanoi-trained Khmer dictated their treatment. As these Cambodia Communists returned, they were disarmed and assigned low-ranking positions. Their duties largely included the training of Khmer Rouge recruits and in this way, they were crucial in establishing a solid foundation for the CPK. Beginning in 1971, however, these returnees were gradually purged. Large-scale pogroms began around 1973 and up to 1975, the CPK effectively eliminated all Sihanoukist and moderate Communist members of the movement. By 1977, all but a dozen of the 1,000 returnees had been killed by those loyal to Saloth Sar. Counted among the victims were many veteran leaders of the Communist movement. Mey Pho, for example, returned to Cambodia in 1970. In 1945 it was Mey Pho, along

---

88 Isaacs, *Without Honor*, 233; I develop this theme more fully in Chapter 6.

89 Communist Party schools were established, staffed by young men and women who were recruited locally. Cooperative farms were likewise formed as the Khmer Rouge began to implement its population resettlement programs. "Liberated" towns and villages were evacuated, with the inhabitants forced to labor in work-brigades. Other programs initiated included the suppression of Buddhism, the introduction of dress codes, and the formation of youth groups. Detention and interrogation (i.e., torture) centers were also established.

90 Chandler, *Brother Number One*, 91.

91 Chandler, *Tragedy of Cambodian History*, 210.

with six other men, who attempted to overthrow Sihanouk. He had spent 14 years studying Marxism-Leninism in Vietnam and two years in China. After the Khmer Rouge came to power, however, he was arrested and killed. Other notable veteran leaders include Keo Moni, Leav Keo Moni, Sos Man, and Yun Soeun.[92]

The Death of Cambodia

> *... sometimes the bombs fell and hit little children, and their fathers would be all for the Khmer Rouge ...*
>
> Chhit Do, CPK leader near Angkor Wat[93]

With Sihanouk's fall, the Vietnam War, in Arnold Isaac's words, "fell on his helpless country like a collapsing brick wall."[94] American bombing intensified, spreading fear throughout the countryside. Brutal fighting between Lon Nol's forces, the Vietnamese Communists, and the Khmer Rouge wrought a terrible toll on Cambodia's population in terms of death and displacement. Phnom Penh and other provincial towns swelled in size, as hundreds of refugees sought safety in the cities. Isaac explains that disruption of civilian life was virtually complete. Indeed, the "killing" of Cambodia generated a new Khmer expression: "The land is broken."[95] Isaacs, who witnessed the final years of the war as a correspondent for the Baltimore *Sun*, describes one journey he took outside Phnom Penh in 1973. "Once, pleasant villages had stood almost shoulder-to-shoulder on that highway," he writes. "Now they were not just ruined, but obliterated. For five miles, not a house still stood on either side of the road, or as far away as one could see across the fields. No trees were left, just broken stumps. In a few places grass had begun to grow again, but most of the land was blackened and dead. It was as if the [B-52] bombers had sought to destroy the earth itself."[96]

And the expanding war proved devastating to Cambodia's land and people. Within two years approximately one-third of all of the country's bridges were destroyed, two-fifths of the road network was made unusable, and the railroad was out of operation. Much of the country's infrastructure, including its lone oil refinery near Kompong Som, had stopped functioning. Only 300 of 1,400 rice mills and 60 of 240 saw mills were in operation and both timber and rubber production—Cambodia's major pre-war commercial products—had declined to only one-fifth of prewar production levels. Moreover, upwards of half of Cambodia's livestock had been killed, either through fighting, bombing, or as a food source for the starving people.[97]

---

92 Son Ngoc Minh died mysteriously in 1972 while visiting China for medical treatment.

93 Quoted in Kiernan, *The Pol Pot Regime*, 23.

94 Isaacs, *Without Honor*, 199.

95 Isaacs, *Without Honor*, 209.

96 Isaacs, *Without Honor*, 218.

97 Isaacs, *Without Honor*, 224.

Cambodia's economy was in ruin, shattered by misplaced foreign loans and governmental corruption. According to Shawcross, the United States had begun selling to Cambodia surplus American agricultural products—wheat, flour, vegetable oil, tobacco, cotton fiber and cotton yard—under the "Food for Peace" program. These commodities were purchased with Cambodian riels, which were then placed in a blocked account in Phnom Penh and used to pay the salaries of the expanding army. The "sale" of American agricultural produce thus financed a new military machine.[98] And this military machine was riven with corruption. Officers within Lon Nol's military padded their manpower lists with "ghost" soldiers, people who existed in name only. Estimates placed the number of phantom soldiers somewhere between 40,000 and 80,000, representing up to $2 million a month that ended up in the pockets of corrupt officers. And the landscape of Phnom Penh was a visual reminder of the uncontrolled corruption. Haggard and orphaned children shared the streets of the capital with newly purchased Mercedes, Peugeots, Audis, and other luxury cars; starving and traumatized refugees begged in front of newly constructed villas that housed the corrupt generals and politicians.[99]

The people of Cambodia suffered from both absolute and relative food shortages. At times, because of corruption, inefficiency, or the vagaries of warfare, there was rarely enough food. Other times, food was present, but most people could not afford it because of hyperinflation, exorbitant prices, and declining real wages. American policies, likewise, directly led to malnutrition, starvation, and death. The United States, for example, consistently kept rice imports to a bare minimum, supposedly to prevent "food riots" among the starving people and to "encourage" greater domestic rice production. How the Cambodian peasantry, under constant American B-52 bombardment, were to harvest rice was not specified. The proximate reason for the lack of food imported into Cambodia, however, is the political spin-maneuvering of the Nixon administration. As the months dragged into years, and the plight of Cambodia's people steadily deteriorated, American officials refused to acknowledge a crisis. Talk of a "refugee problem" was drowned out by the rhetoric of an American and Lon Nol victory, and the salvation of the world from communism. Simply put, as Shawcross concludes, by keeping rice imports low, Washington was able to disguise the serious nature of the problem, namely that the Cambodian people were starving to death. Issacs is equally blunt in his assessment: "Long after it was apparent to everyone else that civilian suffering was acute and growing worse, and that the Khmer government was neither competent enough nor concerned enough to do anything about it, American officials continued to deny any need for a major relief effort. From the start of the war almost to its end, instead of food and medicine the US government supplied only a long list of statements declaring, in absolute contradiction to all the evidence, that Cambodia's refugees were being adequately cared for."[100]

Conditions for the sick and wounded—increasing daily as the war intensified—likewise deteriorated. Of the 29 civilian hospitals that existed prior to the onset of war

---

98 Shawcross, *Sideshow*, 221.

99 Isaacs, *Without Honor*, 208–210; Shawcross, *Sideshow*, 221.

100 Shawcross, *Sideshow*, 222–223; Isaacs, *Without Honor*, 220..

in 1970, only 13 were operating a year later; the rest had been destroyed by ground-fighting or by carpet bombing campaigns. And of those hospitals remaining, space and supplies were desperately short. Conditions were deplorable as patients slept on cots, rush mats, wooden benches, or the floor. Doctors and nurses complained to American officials of a lack of antibiotics, vitamins, antimalarial drugs, dressing materials, and surgery equipment. By 1971 hundreds of people were dying for want of proper treatment and public health specialists and relief agencies were reporting rising malnutrition and vitamin deficiency. Shawcross describes one Phnom Penh hospital, where fifteen percent of all infants were dying of an easily controlled gastric disorder because of a lack of appropriate medicine.[101]

Again, American attitudes and policies contributed to the death of Cambodia. In 1969, before the war, Cambodia had imported $7.8 million worth of drugs, paid for by exports of rice and rubber. By 1971, with demand for medicine and other supplies increasing at an astronomical rate, Cambodia was only able to import $4.1 million worth of material. More pressing was the fact that pharmaceuticals supplies were *not* included under the US AID Commodity Import Program. Faced with widespread shortages of essential medicines, Cambodian officials repeatedly asked American officials to include these drugs on the list of permissible imports. And repeatedly, American officials declined. In part, the refusal on the part of the American advisors was to mask the refugee problem. The official reason given, however, was that the Cambodians did not properly know how to use the medicines and that the drugs would be sold to the enemy. While there is some truth to the second reason—the result, of course, of the uncontrolled inflation that gripped the capital—the first reason epitomizes the callousness and condescension exhibited by many American officials who dictated Cambodian policy.[102] It was Ambassador Emory Swank, after all, who told investigators from the US General Accounting Office that it was "the policy of the United States not to become involved with the problem of civilian war victims."[103]

Ironically, peace in Vietnam spelled tragedy in Cambodia. In January 1973 the United States and the Democratic Republic of Vietnam agreed to an agreement to end the war. As part of the conditions of the Paris Peace Accords, all parties were to respect Cambodia's independence, sovereignty, territorial integrity, and neutrality. The treaty also required all foreign countries to end their military activities in Cambodia and to withdraw their troops; foreign countries, likewise, were not to use Cambodia to encroach on the sovereignty or security of other countries. Consequently, American officials encouraged Lon Nol to declare a unilateral cease-fire and to insist on a Vietnamese withdrawal from Cambodia. Soon thereafter Khieu Samphan and Penn Nouth signed a statement welcoming the cease-fire in Vietnam and asserting that they too cherished peace.[104]

However, as Isaacs explains, if the United States gave little thought to Cambodia's needs in deciding to make war in Indochina, the United States was no more mindful

---

101 Shawcross, *Sideshow*, 225; Isaacs, *Without Honor*, 209.
102 Shawcross, *Sideshow*, 225; Isaacs, *Without Honor*, 220–221.
103 Quoted in Isaacs, *Without Honor*, 221.
104 Clymer, *Troubled Relations*, 130.

of Cambodia when it sought to achieve peace in Indochina. US negotiators read a unilateral statement into the record, telling the representatives of the DRV that the Lon Nol government would halt offensive operations as soon as the Vietnamese cease-fire took effect. Consequently, a *de facto* truce in Cambodia could result if the insurgents reciprocated. Conversely, if the Communists continued to attack, Lon Nol's forces and the United States would take "necessary counter-measures" and the US "would continue to carry out air strikes in Cambodia as necessary until such time as a cease-fire could be brought into effect."[105]

Whereas both the DRV and the US paused to see what would transpire of the truce, the Khmer Rouge were not so inclined. For the CPK leadership, the Paris Accords were seen as the final betrayal. Having come so far, and with victory in sight, Saloth Sar and his colleagues were unwilling to revert back to a political struggle; nor were they amenable to sharing political power with Lon Nol's government. The Khmer Rouge vowed to continue the fight.[106] In February Khmer Rouge forces attacked Kompong Thom; other attacks were launched throughout the countryside, as the Cambodian Communists moved to extend their territorial control.

Nixon and Kissinger, both falsely claiming that the Vietnamese had broken the cease-fire, resumed the horrendous aerial attacks on Cambodia. Beginning on February 9, and continuing for six months, the United States launched an air war "with unprecedented fury."[107] Despite claims to the contrary, the bombing campaign was unjustified and illegal. The United States had no troops in Cambodia and had never official recognized the Lon Nol government as an ally. Furthermore, the US military actions were in contravention of the Paris Peace Accords. However, the war in Cambodia was, in the words of CIA director William Colby, "the only game in town."[108]

America's bombing campaign of the spring and summer of 1973 was the most brutal and intense display of firepower yet. In March 1973 American B-52s dropped over 24,000 tons of bombs on Cambodia. In all of 1972 "only" a total of 37,000 tons had been dropped. By April, the tonnage increased to 35,000 and in May the figures topped 36,000. Bombing runs had become so prevalent that air traffic congestion became a major problem. According to Shawcross, the "bombing of Cambodia was now so intense that the Seventh Air Force was faced with serious logistical problems. At one stage B-52 sortie rates were as high was 81 per day. In Vietnam the maximum had been 60 per day. The Seventh Air Force history for the period notes that, with the Cambodian sky so crowded, the problems of air-traffic congestion were considerable; sorties were so frequent that it was impossible to given adequate 'Air Strike Warnings' to other aircraft."[109] America's logistical trials and tribulations, however, were borne disproportionately by the Cambodian people.

---

105 Isaacs, *Without Honor*, 211–212.

106 Chandler, *Brother Number One*, 95.

107 Isaacs, *Without Honor*, 217.

108 Quoted in Chandler, *Tragedy of Cambodian History*, 225; for an in-depth discussion on the illegality of the air campaign, see Isaacs, *Without Honor*, 226–228, Shawcross, *Sideshow*, 280–310, and Clymer, *Troubled Relations*, 137–139.

109 Shawcross, *Sideshow*, 294–95.

In one well-documented case, the ferry port of Neak Luong, located thirty-eight miles southeast of Phnom Penh on the Mekong River, was accidentally bombed when a B-52 crewman forgot to flip a switch that would direct the aircraft to its bomb-release point. The plane's entire load was dropped on the center of town, killing 137 people and wounding another 268. Most of the central market and city hospital were destroyed.[110]

By the time the air campaign was concluded in late 1973, American B-52s had dropped over 260,000 tons of bombs on Cambodia—a figure that does *not* include the tonnage dropped by other American fighter planes. In comparison, the Allies dropped a total of 160,000 bombs on Japan during the *entire* course of the Second World War. As Chandler concludes, this tonnage was "dropped on a country that was not at war with the United States and that had no US combat personnel within its borders."[111] Estimates of Cambodian casualties range from 150,000 to nearly 750,000.[112]

The bombing was not only illegal, it seemed inexplicable. As Isaacs writes, "Through it all, as if the US could think of nothing else to do, the bombing grew in fury."[113] The air campaign was a brutal, senseless campaign that inflicted a massive loss of life on Cambodia and further devastated an already deteriorating infrastructure. Kamm likewise writes that "With the callous disregard of the interests of the Cambodian people that marked all of America's wartime involvement in that country, and in full knowledge that Cambodia's demented and corrupt regime could only prolong their people's suffering, America did all that it could to drag out senselessly the life of a hated government and a war that Washington knew was lost."[114] But then again, neither the bombing of Cambodia, nor the war itself, had anything to do with Cambodia. Rather, Cambodia—and for that matter all of Indochina—was a surrogate space, an arena in which the United States sacrificed the Cambodians, the Vietnamese, the Laotians in an attempt to expand their global economic reach.[115] Short concludes that for the United States following the Paris Peace Accords, "Cambodia was the one place in Indochina where it could flex its

---

110 Isaacs, *Without Honor*, 229.

111 Chandler, *Brother Number One*, 96.

112 Tully, *Short History*, 167.

113 Isaacs, *Without Honor*, 225.

114 Kamm, *Cambodia*, 116.

115 For an in-depth discussion of Vietnam, and Southeast Asia, as a surrogate space in the building of the American empire, see James A. Tyner, *America's Strategy in Southeast Asia: From the Cold War to the Terror War* (Boulder, CO: Rowman & Littlefield, 2007). Nothing indicates the sacrifice of Cambodia by American officials more than Kissinger's statement to the Chinese Prime Minister, Zhou Enlai. Privately, in 1973, Kissinger conceded that Cambodia was doomed, that Communist forces would emerge victorious, and that the Lon Nol regime would collapse. However, he persisted in championing the devastating aerial attacks on Cambodia for another two years. In November 1973 Zhou Enlai asked why the Americans continued to support Lon Nol. "I will speak frankly," Kissinger replied. "Our major problem with Cambodia is that the opponents of President Nixon want to use it as an example of the bankruptcy of his whole policy. So if there is a very rapid collapse, it will be reflected in our other policies. That," he said, "is our only concern." As Clymer concludes, several thousand

military muscle and show that, even in retreat, it was still capable of something. Bombing became a virility symbol."[116]

Aside from the devastating toll on Cambodia's people and land, it was American military intervention in 1973 that more than anything paved the way for the Khmer Rouge to rise to power.[117] Kiernan is blunt in his assessment: "Pol Pot's revolution would not have won power without [the] US economic and military destabilization of Cambodia."[118] This was known at the time. Throughout 1973 American and Khmer Republic intelligence reports indicated that the aerial bombardments caused massive civilian losses and that the survivors were turning to the CPK for support.[119] On May 2, 1973, for example, the CIA's Directorate of Operations provided details on a new recruiting drive launched by the CPK:

> They [the CPK] are using damage caused by B-52 strikes as the main theme of their propaganda. The cadre tell the people that the Government of Lon Nol has requested the airstrikes and is reponsible for the damage and the 'suffering of innocent villagers' ... The only way to stop 'the massive destruction of the country' is to ... defeat Lon Nol and

---

more Cambodians would have to die because Nixon's political "opponents" in the United States might benefit if the war ended too soon. See Clymer, *Troubled Relations*, 146.

116 Short, *Pol Pot*, 245.

117 Some US officials did acknowledge both the deteriorating conditions and the fact that American military actions were in fact buttressing the growth of the Khmer Rouge, and that the Khmer Rouge were not mere puppets of the DRV. Most of these voices, however, were silenced within the Nixon administration. As a case in point, in one document prepared by the State Department in January 1973 for Vice President Spiro Agnew, the above analyses were spelled out. However, prior to Agnew receiving the document, Kissinger modified the interpretations. According to Kissinger the Vietnamese Communists continued to control the indigenous Communist military and political apparatus in Cambodia—even as late as 1975, Kissinger refused to accept that the Khmer Rouge were not dependent on Hanoi—and while Kissinger acknowledged that the enemy controlled most of eastern and northeastern Cambodia, he maintained that Lon Nol's government forces controlled a substantial part of the country, including the richest rice producing regions and the majority of the population. See Shawcross, *Sideshow*, 126–27.

118 Kiernan, *The Pol Pot Regime*, 16. John Shaw sanctimoniously writes that the Cambodian people were the "victim's of Hanoi's willingness to sacrifice their welfare for its conquest of the South." He maintains that Operation MENU, the subsequent US and ARVN incursions, and the American bombing campaign of 1973 were "tragic military necessities, driven by self-defense against Hanoi's aggression." He concludes that "Any assessment of external influences on recent Cambodian history [meaning the subsequent genocide] must lay primary responsibility at Hanoi's door." See John M. Shaw, *The Cambodian Campaign: The 1970 Offensive and America's Vietnam War* (Lawrence: University of Kansas Press, 2005), 169. Shaw, however, fails to consider the split between the Vietnamese Communist Party and the Cambodian Communist Party. More troubling, Shaw fails to consider the larger context of the First and Second Indochina Wars, namely the fact that the war in Vietnam was an anti-colonial war and not, as he indicates, a *civil war* between a sovereign DRV and a sovereign RVN. Consequently, he fails to consider the necessity of US involvement in either Vietnam or Cambodia. Despite Shaw's revisionist history, the 1973 bombing of Cambodia was tragic because it was *unnecessary* from either a political or military standpoint.

119 Kiernan, *Pol Pot Regime*, 20.

stop the bombing. This approach has resulted in the successful recruitment of a number of young men ... Residents ... say that the propaganda campaign has been effective with refugees and in areas ... which have been subject to B-52 strikes.[120]

The bombing did not forestall defeat of Lon Nol's anti-communist government in Cambodia, nor did the bombing prevent the emergence of a Communist force from assuming power. Rather, American military actions played into the hands of the Khmer Rouge. While the bombings ended in mid-August of 1973, horrendous fighting continued to be waged across Cambodia's landscape for another nineteen months. Isaacs relates that the Khmer Rouge pressed toward Phnom Penh with a profligate, disastrous disregard for their own losses. From 1973 onward, Cambodia, to Isaacs, seemed "an entire country gone amok."[121]

By 1974 Phnom Penh was surrounded and experienced daily shelling. The airport was heavily damaged by mortar attacks, and the Mekong was mined. Isolated, and under siege, the people of Cambodia's capital city—most of whom were peasants who had fled the war in the country—endured even greater hardships. Little foreign relief was forthcoming. American officials did provide some assistance through airlifts of food and fuel to Phnom Penh between February and April. However, as Shawcross explains, the airlift of food prevented famine, but it did not stop starvation from spreading. Relief organizations and investigations by the World Health Organization, and by the Senate Refugee Subcommittee, identified "simply put" that the Cambodians—and especially the children—were starving. Shawcross describes the conditions:

> In the camps and in the streets, in the cardboard shelters, in the Cambodiana Hotel refugee center, one could see sick children everywhere. Those who suffered from kwashiorkor, extreme protein deficiency, had distended bellies and swollen hands, feet and ankles. Their hair was falling out or turning light brown, and so was their skin; they behaved as listlessly as one might expect. Other children had simply far too little to eat to be able to grow properly and were suffering from marasmus. Their matchlike limbs hung over the empty skin folds of their bodies, they had almost no muscular control, and eight year-olds looked like shriveled babies.[122]

According to Isaacs, as late as March 1975, six weeks before the end of the war and while Cambodian children were sick from disease and malnutrition, and were starving to death in the hospitals and pagodas of Phnom Penh, the embassy officer in charge of economic assistance programs explained of the efforts to help the refugees: "We are trying to assess the magnitude of the problem. None of the agencies with which we're working has a definite fix ... We'll have a better feel for it in a couple of weeks."[123] The problem was, Cambodia's people did not have time.

On April 17, 1975 Cambodia died.

---

120 Quoted in Kiernan, *Pol Pot Regime*, 22.
121 Isaacs, *Without Honor*, 230–232.
122 Shawcross, *Sideshow*, 348.
123 Isaacs, *Without Honor*, 224.

Chapter 4

The Un-Making of Space

A revolution that does not produce a new space has not realized its full potential ...

Henri Lefebvre[1]

Destroy the old order, replace it with the new.

[Khmer Rouge slogan][2]

Karl Marx, in his voluminous writings, proposed a scientific understanding of the history of humankind. He believed that history is a necessary *process*, one that foreshadows an inevitable culmination in Communism. Marx also understood, however, that history is propelled through a series of revolutions. Each new era, advancing dialectically, is transformed through revolutionary praxis.

Saloth Sar and his colleagues deliberately set Cambodia on a revolutionary path toward—in their minds—a Communist utopia. As David Chandler explains, they explicitly sought to refashion, or remake, Cambodia. However, unlike Marx's scientific reasoning, the Khmer Communist's decision to "wage revolution everywhere in Cambodia did not spring from a study of Cambodian social conditions or from consultation with others, but from a conviction on the part of the CPK's [Communist Party of Kampuchea] leaders that a recognizably communist revolution needed to be waged. If the right preconditions did not exist, that problem could be overcome by revolutionary fervor."[3] Saloth Sar, however, had apparently not learned a key insight offered by Marx, namely that Marx "knew he could not impose his own will on the course of history." Rather, Marx "condemned conspiratorial revolutionaries who wished to capture power and introduce socialism before the economic base of society had developed to the point at which the working class as a whole is ready to participate in the revolution."[4]

The Khmer Rouge, in opposition to classical Marxist thought, believed in a doctrine of *their own inevitability*. For the Khmer Rouge, the wheels of history turned only for themselves. According to a widely spoken expression during the reign of the Khmer Rouge, "The wheel of history is inexorably turning: he who cannot keep

---

1 Henri Lefebvre, *The Production of Space*, translated by Donald Nicholson-Smith (Oxford, UK: Blackwell, 1991 [1974]), 54.

2 Quoted in Henri Locard, *Pol Pot's Little Red Book: The Sayings of Angkar* (Chiang Mai, Thailand: Silkworm Books, 2004), 271.

3 David P. Chandler, *The Tragedy of Cambodian History: Politics, War, and Revolution since 1945* (New Haven, CT: Yale University Press, 1991), 239.

4 Peter Singer, *Marx: A Very Short Introduction* (Oxford, UK: Oxford University Press, 2000), 79.

pace with it shall be crushed."[5] As Henri Locard explains, this reasoning seemed to confirm a belief prevalent among the disciples of Marx that they had found "the true meaning of the history of mankind: the glorious (if not bloody) march towards a Communist society." Furthermore, since "all the revolutionaries who fought for the proletarian revolution were engaged in a good cause, all means for attaining it were good."[6] In other words, those events that transpired during and after the revolution were considered "just."

On January 16, 1977, Nuon Chea spoke before assembled audience in Phnom Penh. Nuon Chea was a long-time revolutionary. He attended university in Bangkok and it was in Thailand, in 1950, that he joined the Thai Communist Party. He later transferred his membership to the Vietnamese-dominated Indochinese Communist Party (ICP). After the Communist Party of Kampuchea (CPK) was established in 1960, Nuon became deputy secretary of its Central committee and a member of its Standing Committee; these were the two most senior bodies responsible for Party policy.[7] And it was in his position as the second-most powerful cadre in the CPK that Nuon spoke in 1977, commemorating the ninth anniversary of the Revolutionary Army of Democratic Kampuchea. He explained:

> We liberated our country on 17 April 1975 … We did that for the defense of Democratic Kampuchea, for the Cambodian workers and peasants in cooperatives, for the next decade, the next century, the next millennium, the next ten thousand years, and forever.[8]

Drawing heavily, but unacknowledged, on the sentiments of Mao Zedong, Nuon Chea spoke to the temporal dimensions of the revolution; he spoke of the *history* of a new Cambodia. However, as this chapter demonstrates, Nuon Chea—along with Pol Pot and the other high-ranking Khmer officials—knew very well that a new Cambodia required also a new space. They put into practice what the French scholar Henri Lefebvre put into words: "A revolution that does not produce a new space has not realized its full potential." Otherwise, as Lefebvre continues, such a revolution "has failed in that it has not changed life itself, but has merely changed ideological superstructures, institutions or political apparatuses." According to Lefebvre, "A social transformation, to be truly revolutionary in character, must manifest a creative capacity in its effects on daily life, on language and on space."[9] And therein lies the basis of another widely-spoken saying of the Khmer Rouge: "To build the new society, there must be some new people."[10]

---

5 Locard, *Pol Pot's Little Red Book*, 213.

6 Locard, *Pol Pot's Little Red Book*, 211–212.

7 Stephen Heder and Brian D. Tittemore, *Seven Candidates for Prosecution: Accountability for the Crimes of the Khmer Rouge*, 2nd edition (Phnom Penh: Documentation Center of Cambodia, 2004), 59.

8 Locard, *Pol Pot's Little Red Book*, 212.

9 Lefebvre, *Production of Space*, 54.

10 Locard, *Pol Pot's Little Red Book*, 293.

Cambodia and the Meaning of Revolution

The word *revolution* has two dominant meanings. On the one hand, *revolution* (or *political revolution*) refers to any and all instances in which a state or political regime is overthrown and thereby transformed by a popular movement.[11] According to this usage, Cambodia (as detailed in Chapters 2 and 3) reveals a series of revolutions. The Khmer Issaraks, the Viet Minh-trained Khmer resistance groups, as well as the Paris-educated Khmer Rouge all initiated revolutions. Targets were variously identified as the French colonial authorities, the monarchy (as embodied by Sihanouk), the Lon Nol/Sirik Matak government, and the United States. The formation of a unified revolutionary movement in Cambodia, however, was always precarious. Although a common interest revolved around a basic desire to rid the country of French dominance (an anti-colonial movement), revolutionaries were always divided in terms of ideological approach. The Communist movement, for example, did not initially attract much popular support. It was not until the devastation wrought by the American bombing and the removal of Sihanouk by Lon Nol that people rallied in large numbers to the Khmer Rouge. This is a point that bears emphasis, because the Khmer Rouge never felt that they represented the interests of the population. And in fact they never did.

On the other hand, a more restrictive usage defines revolutions as including "not only mass mobilization and regime change, but also more or less rapid and fundamental social, economic, and/or cultural change during or soon after the struggle for state power." For these revolutions, Goodwin suggests the term *social revolution*.[12] This definition resonates with Colburn's assertion that the "purest meaning of revolution is the sudden, violent, and drastic substitution of one group governing a territorial political entity for another group formerly excluded from the government, *and* an ensuing assault on state and society for the purpose of radically transforming society."[13] Social revolutions, therefore, include two components: the actual change in governance, and the transformation of society. More than a political revolution, what the Khmer Rouge imagined in Cambodia was a social revolution.

David Chandler suggests that the Cambodian revolution was a courageous yet doomed attempt by a group of utopian thinkers to break free from the capitalist world system, abandon the past, and to rearrange the future.[14] This sentiment is captured in a highly optimistic Khmer Rouge planning document: "Beginning in 1980, the

---

11 Jeff Goodwin, *No Other Way Out: States and Revolutionary Movements, 1945–1991* (Cambridge: Cambridge University Press, 2001), 9.

12 Goodwin, *No Other Way Out*, 9.

13 Forrest D. Colburn, *The Vogue of Revolution in Poor Countries* (Princeton, NJ: Princeton University Press, 1994), 6. Robert Melson, likewise, argues that a "social revolution is a fundamental transformation, usually carried out by violence, in society's political, economic, and social structures and cultural values and beliefs, including its reigning ideology, political myth, and identity." See Robert Melson, *Revolution and Genocide: On the Origins of the Armenian Genocide and the Holocaust* (Chicago: University of Chicago Press, 1992), 32.

14 David P. Chandler, *Brother Number One: A Political Biography*, revised edition (Chiang Mai, Thailand: Silkworm Press, 2000), 3. Chandler also asserts that the revolution sprang from a colossal misreading of Cambodia's political capacities, its freedom of maneuver,

Angkar will create a model society that exists nowhere else in the world, where everyone will eat his fill three times a day, where everyone will live well, in villages and towns, where there will be no more classes."[15]

The social revolution of the Khmer Rouge, however, was not to be widely publicized. Indeed, Serge Thion maintains that the most striking feature of the idea of revolution entertained by the Khmer Communists was that it was *unexpressed.* He explains that revolution "was never a major propaganda theme because peasants could not be convinced by, and would have shied away from, any involvement in a complete overturning of their world (and cosmic) order." Consequently, in the early months of Democratic Kampuchea, the "revolution and the existence of a revolutionary party were not only played down in propaganda" but were "revealed only to the enlightened few who could achieve senior positions in the apparatus."[16] This is a crucial observation, for it directs our attention to the *geographical imaginations* expressed by only a few, select revolutionaries. The communist revolution engineered by the Khmer Rouge was neither a populist uprising nor a spontaneous grass-roots movement—although, certainly, these elements were present. Rather, the ascension of Saloth Sar and his colleagues represents an elitist movement that coopted the grievances of the masses. Consequently, the experiences of the Khmer Rouge support the argument that while the origins of contemporary revolutions are rooted in social, political, and economic conflict, the outcomes of these revolutions have been determined by the political—*and geographic*—imagination of revolutionary elites. Through their language, images, and daily political activity, revolutionaries strive to reconfigure social and social relations.[17] They consciously seek to break with the past and to establish new societies and new spaces. In the process, they create new social relations and new landscapes. And it is through the actions of these revolutionaries that certain material practices—those efforts to (re)create society and space—are justified and legitimated.

Revolutions provide the context for the implementation of specific geographical imaginations by revolutionaries. Revolutions, according to Melson, "destroy not only the institutions and power of the old regime, they also undermine the legitimacy of the state and place in question the political identity of the community itself. Revolutions, therefore, provide the structural opportunities for ideological vanguards to come to power and to impose their views on society." Furthermore, after these regimes come

---

and the interests of rural poor, coupled with Saloth Sar and his colleague's thirst for power and an unlimited capacity for distrust.

15 Locard, *Pol Pot's Little Red Book*, 293.

16 Serge Thion, "The Cambodian Idea of Revolution," in *Revolution and Its Aftermath in Kampuchea: Eight Essays*, David P. Chandler and Ben Kiernan (eds) (New Haven, CT: Yale University Southeast Asia Studies, Monograph Series No. 25, 1983), 10–33; at 23.

17 Colburn, *Vogue of Revolution*, 5. Somewhat ironically, structural-based theories, and especially those that are Marxist-based, of revolution and genocide are notorious for excluding agency. As Colburn (p. 11) explains, Marxists invariably maintain a faith in using impersonal economic and social structures to explain the origins and outcomes of events. And, ironically, despite Marxist understandings of the role of vanguards, contemporary Marxist scholars continue to attribute radical political change either to the disruption caused by economic growth or to the demands of economic competition.

to power, they attempt to construct a new order that will support the revolutionary state. Melson elaborates that "It is not enough for the revolutionaries to have seized the buildings and offices of the old regime. They desperately need support from society at large, but that is often an inchoate mass itself looking for leadership." Consequently, to implement its geographic vision of a new society and to mobilize support, the revolutionary regime must completely *un-make* all that existed prior to the revolution, and to subsequently construct "a new system of legitimation and to redefine the identity of the political community."[18]

That Saloth Sar and the other members of the Central Committee were influenced by the writings of Marx, Lenin, Stalin, and Mao is not in doubt. The depth of the Khmer leadership's understanding, as well as their particular interpretations, is less clear.[19] It is one thing, for example, to note that Saloth Sar read, say, Marx's *Das Kapital*; it is quite another to know what Sar might have derived from this reading. One wonders how Lenin's statements on national autonomy as opposed to regional autonomy, or self-determination, were translated and understood by the ranking officials of the CPK.

Saloth Sar, in particular, shied away from the public and remained, throughout his life, in the shadows. He purposefully left behind no major works; certainly nothing exists (or has been brought to light) that is comparable to the writings of, for example, Ernesto "Che" Guevara or Frantz Fanon. For Saloth Sar, there is no *Mein Kampf* or *Wretched of the Earth*. Saloth Sar, moreover, explicitly sought to distance the Cambodian revolution from those of China and Vietnam. Accordingly, it was commonplace for the Khmer Rouge to *erase* from the official histories of the CPK any reference to Chinese or Vietnamese influence.

A few insights, provided by the pioneering efforts of David Chandler, Ben Kiernan, Karl Jackson, Serge Thion, and Philip Short (among others) have been established. It is generally agreed, for example, that Saloth Sar's ideas—despite his claims to the contrary—were shaped to a large extent by foreign influences. His French-influenced education was paramount.

In colonial Cambodia, French administrators were slow (or reluctant) to establish an indigenous-based educational system. In the schools of Cambodia, for example, history was taught with no adaptation to local conditions. Khmer children were taught to identify with, and expected to internalize, French history and French political values. This education introduced Cambodian students to *French* ideas and ideals of progress and of the concepts of democracy, imperialism, and revolutionary change.[20] And of their own past—except for that reconstructed and repackaged by the French—this was almost completely ignored. World history, as Thion explains, "was an obscure struggle, peopled with the likes of Caesar, Napoleon, and Bismark

---

18 Melson, *Revolution and Genocide*, 18.

19 See the discussions in Philip Short, *Pol Pot: Anatomy of a Nightmare* (New York: Henry Holt and Company, 2004) and Chandler, *Brother Number One*.

20 Chandler, *Brother Number One*, 6.

fighting one another in a vast rice field dotted with sugar palms."[21] French, simply put, was "the prism through which [Cambodian children] viewed the outside world."[22]

Many of the Khmer Rouge leadership were part of a privileged few Khmer who even had access to education. As of 1947, for example, only a few thousand other Cambodians had progressed as far as Saloth Sar had in education. And even though Saloth Sar failed many of his examinations, one should not forget that he was in a position to sit the examinations, something most of Cambodia's peasants could not even dream of. Moreover, Saloth Sar and his associates came from relatively wealthy families. They neither farmed the land nor gathered fish as most Cambodians did. As Chandler concludes, Saloth Sar's conversion to political activism and socialist ideals—similar to that Khieu Samphan, Ieng Sary, and others—occurred against this privileged background and before he had any experience of working for a living.[23]

Two groups of Khmer revolutionary nationalists existed in Cambodia: those who were influenced by, and trained with, the Vietnamese, and those who studied in France. During the late 1940s and early 1950s, only about one hundred Cambodians were pursuing advanced studies in France. Most of these students were politically neutral, although a few were increasingly influenced by the radical politics and anti-colonial ideas in circulation at the time. As Thion identifies, those who went to France were exposed to a more diverse political experience, an experience that would remarkably affect their geographical imaginations.[24] Many of these radicalized students, for example, including Saloth Sar, Ieng Sary, Son Sen, Khieu Samphan, Hou Youn, and Hu Nim, would join other organizations, such as the Communist Party of France (CPF).[25]

Saloth Sar arrived in Paris on October 1, 1949 and soon joined the Khmer Student Association (AEK; *l'Association des Estudiants Khmers*). According to Chandler, most Cambodian students in France saw themselves as intellectuals and, as such, it was but a small step for them to see themselves as a vanguard, ahead and above the rest of their compatriots.[26] This is perhaps most clear in the establishment of informal discussion groups that many Khmer students participated in. Through the AEK, a number of discussion groups, or "Study Circles," were initiated, known as *Cercles d'Etudes*. There was a "Law Circle" and an "Arts Circle," for example, and other groups addressed literature, farming, or women's issues. There was another, more secretive and unnamed group, however, that met to discuss political issues and, in particular, Communist texts. This group was led by two highly influential Khmer students: Keng Vannsak and Ieng Sary. Vannsak, a study of Cambodian philology, had arrived in Paris in 1946. Given his seniority, he often played the role of "older brother" for the younger Cambodian students. As a linguist and scholar of Cambodian literature, he attempted through his studies to uncover the pre-Buddhist, pre-Sanskrit layers of Cambodian vocabulary and hence culture. He also promoted a radical idea

21 Thion, "Cambodian Idea of Revolution," 14–15.
22 Short, *Pol Pot*, 47.
23 Chandler, *Brother Number One*, 21.
24 Thion, "Camobdian Idea of Revolution," 19.
25 Chandler, *Tragedy of Cambodian History*, 52.
26 Chandler, *Tragedy of Cambodian History*, 52.

of Khmer self-determination and self-reliance, a position that held that Cambodia could—and should—be cut off from other cultures.[27] Ieng Sary was a close associate of Vannsak. He was also decidedly more radical. Of Chinese-Khmer heritage, Ieng Sary was born in southern Vietnam and attended the elite *Lycèe Sisowath* in Phnom Penh. He later won a government scholarship to study in France in 1950, where he took courses in commerce and politics at the *Institut d'Etudes*. Active in strikes and other political acts, Ieng Sary was already relatively fluent in Marxist thought.[28]

And so it was that a small group of Khmer students met two or three times each month in Vannsak's apartment. Some students participated because they were members of the CPF; others attended because their friends participated. And it was through this group that Ieng Sary introduced Vannsak to Saloth Sar. To this point, Saloth Sar apparently spent most of his time reading. According to Chandler's research, Sar was drawn to nineteenth-century French poetry—Hugo, Rimbaud, Verlaine, and Vigny—though his favorite author was said to be the eighteenth-century philosopher Jean-Jacques Rousseau. Chandler concludes, however, that "It would probably be misleading ... to endow his activities at this stage with too much coherence or ambition."[29]

According to Chandler, the Marxist group discussed Stalin's "The National Question" and Lenin's "On Imperialism." They also read, in parts if not in its totality, Marx's *Das Kapital*. Students were exposed to commentaries and articles that appeared in the various journals produced by the French Communist Party, for example *Humanities* and *Les Cahiers de Communisme*. Usually, the works, or extracts, of Marx, Lenin, or Stalin were read in French; the discussions that followed were conducted in a mixture of Khmer and French.[30]

Another key event occurred in August 1951. During that month Thiounn Mumm, another influential Khmer student, traveled to East Berlin with a handful of other members of the Marxist study group. The occasion of their journey was to participate at the International Youth Congress, a remarkable event that was sponsored by the Soviet Union. More than 100,000 delegates from all over the world were in

27 Chandler, *Tragedy of Cambodian History*, 53.

28 Chandler, *Tragedy of Cambodian History*, 53.

29 Chandler, *Brother Number One*, 32. Chandler also indicates that it is unknown what other writers influenced Sar's thinking. Aside from a brief reference to Mao Zedong in his writings, Saloth Sar never mentioned any specific writer.

30 Chandler, *Brother Number One*, 32–33. It is interesting to speculate what other writings Saloth Sar might have encountered. The Paris of the late 1940s and early 1950s was dominated by the daunting figures of Albert Camus, Maurice Merleau-Ponty, and Jean-Paul Sartre. Other intellectuals, on the ascension, included Michel Foucault, Henri Lefebvre, and Roland Barthes. As will become clear in this and subsequent chapters, I am struck by the remarkable, yet perhaps not surprising, parallels between certain facets of the Khmer Rouge and the writings of these French philosophers and social theories. Certainly both Lefebvre and Foucault, for example, were active in Marxist circles, and their ideas would have been widely discussed in various forums, journals, and so forth. An important task remains to connect the earlier writings of these theorists with those of the Khmer Rouge. Moreover, an additional task worth pursuing is the influence that Fanon, in particular, and the Algerian Revolution in general, might have had on the Khmer Rouge leadership.

attendance, and it was here that the Cambodian students first learned of the Khmer People's Revolutionary Party (KPRP) and the Communist-influenced resistance movement that was underway in Cambodia.[31]

Through education and exposure to other Communist organizations, Saloth Sar and other Paris-educated Khmer students began to develop their own ideas regarding Marxism and revolution. These ideas, as discussed in Chapters 2 and 3, began to harden into an anti-colonial and anti-monarchical position. Furthermore, a strong nationalist sentiment permeated their political thinking, an attitude that would culminate in an extreme belief in self-determination and self-reliance. But what, specifically, was in the writings of Marx and his followers that would culminate in the mass violence inflicted upon Cambodia? Where in Marxist doctrine did Saloth Sar and his cadre find the justification for their killing of Cambodia? These questions require a closer examination of the Marxist teaching of revolution.

Marx and the Idea of Revolution

Nineteenth-century Europe, for Karl Marx, reflected both the heights and depths of humanity.[32] On the one hand, Marx recognized and appreciated that the Industrial Revolution—that momentous shift from feudalism to capitalism—had created previously unimagined levels of production. On the other hand, despite the wealth that was being generated, more and more people toiled and lived in inhumane conditions. Marx was driven by the observation that for the first time in human history, humanity had created the means of production that could, in principle, provide for all people, liberating them from compulsive toil; and yet that very same economic system used to industrialize production distributed its bounty to a few wealthy people, thus artificially perpetuating the enslavement of the masses.[33] Furthermore, this state of affairs, the rampant social inequality and injustices that typified Europe, were increasingly sanctioned and policed by a repressive political environment.[34]

---

31 Chandler, *Brother Number One*, 35. Thion argues that while the KRPR provided the Khmer with a sense of pride and purpose, these students early on demonstrated an antipathy to the perceived "Big Brother" role assumed by the Vietnamese. The Khmer students apparently resented the prominence given the Viet Minh representatives and their Khmer allies, such as Keo Meas, on the Parisian scene and in the international conferences which they attended. See Thion, "Cambodian Idea of Revolution," 20.

32 It is not difficult to imagine the parallels drawn by the Khmer studying in Paris. When reading of the Europe of the nineteenth-century, these students could look to Cambodia, and see a country dominated and exploited by an industrial state (France), and ruthlessly repressed by a monarchy (in the person of Sihanouk). Furthermore, it is reasonable to conclude that the scientific foundation of Marxism would have proved immensely attractive to the Khmer intellectuals studying in France.

33 Leon P. Baradat, *Political Ideologies: Their Origins and Impact*, 8th edition (Upper Saddle River, NJ: Prentice Hall, 2003), 177.

34 Baradat, *Political Ideologies*, 176.

Marx, of course, was not the first to expose the ills of society and of capitalism.[35] As Paul D'Amato explains, "many had done so before him." Marx, however, went further in his condemnation of capitalism, for "he revealed how capitalism developed, how it went into crisis, and how it would meet its end."[36] Simply put, Marx predicted the inevitable demise of capitalism. For Marx, it was an historical fact, one that emerged from reasoned thinking.

Dialectic materialism is foundational to Marxism. It is a concept that stretches back to the ancient Greek philosophers, as modified by later Western philosophers, most notably Georg Hegel. Hegel (1770–1831) advanced a theory in which historical change is determined by dialectic conflict. Eschewing the dominant *principle of noncontradiction*—which held that a concept could not be both itself and its opposite at the same time—Hegel believed that reality is composed of two things: it is both itself and part of what it is becoming. Hegel's belief owes much to the Greek philosopher Heraclitus, who believed that there was no permanent reality except the reality of change. It was Heraclitus, for example, who famously asserted that it is impossible to step into the same river twice. Heraclitus also forwarded the idea that all things contain within them their opposites: death was potential in life, adulthood was potential in childhood. Consequently, for Hegel, it was not unreasonable to surmise that something could be both itself and its opposite.[37]

Leon Baradat provides an illustrative example. Imagine an existing state of affairs as the "thesis." In time, newer ideas or challenges will emerge; these may be termed the "antithesis." Conflict between the thesis and antithesis ensues and the result of this interaction—a dialectic process—will result in a *synthesis* of all the good parts of the thesis and the antithesis. History thus reveals itself as a process of always becoming, moving through a series of struggles between thesis and anti-thesis. For Hegel, such a dialectical process always leads to an improvement, because the negative aspects of the thesis and antithesis were always destroyed. This was a process Hegel termed the "negation of the negative." History was inherently and inevitably progressive, with each new era an improvement over the previous.[38]

Marx adopted the premise of Hegelian dialectics and applied this way of thinking to economic systems. It is important to recognize, as Ingersoll and his colleagues stress, that Hegelian logic is not a logic, but rather an ontology, or an explanation of existence. They elaborate that traditional logic presupposes a fixed, static world where time and motion do not operate; Hegel's dialetic method, conversely, presupposes movement and change, wherein the dialetical process is imbedded in the world of real events. Furthermore, for Hegel (and for Marx), everything in the world is interrelated to every other part of the world; nothing—neither objects nor

35 Marx critiqued capitalism, not because it was inefficient, but because of its consequences. Indeed, Marx appreciated capitalism for its productivity, but he deplored the system because of its is inherent oppression and exploitation. Consequently, Marx advocated that capitalism should (and will be) abandoned for a more equitable system.

36 Paul D'Amato, *The Meaning of Marxism* (Chicago, IL: Haymarket Books, 2006), 10.

37 David E. Ingersoll, Richard K. Matthews, and Andrew Davison, *The Philosophic Roots of Modern Ideology: Liberalism, Communism, Fascism, Islamism*, 3rd edition (Upper Saddle River, NJ: Prentice Hall, 2001), 118–119; Baradat, *Political Ideologies*, 180–181.

38 Baradat, *Political Ideologies*, 181.

ideas—exists independently of, or in isolation from, anything else.[39] Both of these premises would have a direct bearing on the material practices enacted by the Khmer Rouge.

Ontologically, Marxism is consistent with Hegelian logic in that Marx envisioned the dialectic as the fundamental logic of history. However, whereas Hegel assumed that the dialectic was guided by the will of God, Marx surmised that materialism—specifically, social classes—powered the dialectic of history. According to Marx, society's foundation inexorably changes, albeit gradually, causing subsequent societal transformations. Such "movements" were directly related to humanity's relationship with the environment.[40]

More specifically, Marx suggested that history can be seen as the dialectic progression of different modes of production. A mode of production, such as feudalism, capitalism, or socialism, is simply a particular form of economic organization. What most distinguishes one mode of production from another are differences in the means of production (i.e., resources, technology, machinery) and the social relations of production (from which social classes are derived). These latter are determined by the affiliation between people and the means of production, whereby the owners of the means of production benefit the most and thus compose the most influential social group—the ruling class. In an agrarian society, the ruling group would be those who are the greatest landowners; alternatively, in an industrial society, the ruling group would own the factories and equipment.[41]

All societies, for Marx, are composed of two constituent parts: the foundation (or base) and the superstructure. The foundation is material, in that it is composed of both the means of production and the relations of production. Conversely, the superstructure is composed of all nonmaterial institutes, including religion, law, education, and government. According to Marx, the function of the superstructure is to assure that the ruling class maintains its position of dominance within society.[42]

Human history, according to Marx, could be divided into a series of historical stages: *primitive communism*, *slavery*, *feudalism*, and *capitalism*. And furthermore, Marx proposed that each mode of production carries its own seed of destruction. In other words, following Hegelian logic, historical change occurs when the forces of production are impeded by the relations of production. As class antagonisms harden, a revolution is produced which ultimately results in the establishment of a new mode of production.[43] The first era of human history, according to Marx, was based on primitive communism. During this period, there was neither occupational specialization nor divisions of labor. Rather, all peoples participated in productive work and necessarily shared their productivity so that all could survive. However, over time people began to specialize (an antithesis) in the production of certain goods, giving rise to a division of labor. The original collectivity of society disappeared and the concept of private property was born. Different objects were valued differently

---

39 Ingersoll et al., *Philosophic Roots*, 120.
40 Baradat, *Political Ideologies*, 182.
41 Baradat, *Political Ideologies*, 178.
42 Baradat, *Political Ideologies*, 178.
43 Ingersoll et al., *Philosophic Roots*, 133 *passim*.

and the value (or importance) of the individual was based on the things he or she owned and/or produced. Gradually, a series of internal contradictions fostered the emergence of newer forms of economic organization until, in the late eighteenth-century, capitalism emerged as a dominant mode of production.

Capitalism is an economic system in which all economic actors—producers and consumers—depend on the market for their basic needs.[44] Waged labor is a defining characteristic, as is private ownership of the means of production (e.g., capital, land, and labor). Capitalism is driven by certain systemic imperatives, namely those of competition, profit-maximization, and profit-accumulation.[45] Indeed, as David Harvey explains, accumulation is the engine which powers growth under the capitalist mode of production.[46]

Marx recognized that capitalism had increased productivity to the point at which all basic material needs could be satisfied. And yet, he was struck by the widespread exploitation and oppression that accompanied—indeed, was endemic to—capitalism. Exploitation, according to Marx, is derived from the labor theory of value, an idea developed by earlier "liberal" economists, such as Adam Smith, John Locke, and David Ricardo. The labor theory of value is predicated on the *intrinsic* worth of an object. Value is itself a problematic term. One understanding of value, for example, is premised on the exchange value of an object, meaning the amount of money (or payment in kind) for which an item can be exchanged on the market. Conversely, one can speak of sentimental value, use value, or esthetic value. The labor theory of value, however, seeks to provide a standard for measuring intrinsic value, and is based on two types of value that are brought to the production process. These include *constant value* and *variable value*. Constant value includes the cost of the factors of production, such as machinery, resources, and finance. It is said that these factors do not add any value to the item greater than their own intrinsic worth. Labor, conversely, is considered a variable value because it does produce something of greater worth than itself. Therefore, Marx reasoned, the intrinsic value of any object is determined by the amount of labor needed to produce it. Thus, while the price of an object may be determined by market transactions, value is determined by labor time.[47]

For Marx, capitalism enslaves the working class (i.e., the proletariat) because people have to work to survive. Furthermore, given that the capitalist class has a monopoly on the means of production (i.e., landowners, the monarchy), workers are forced to sell their labor on the market. Although this exchange of labor for wages is promoted as a free and open relationship, it in fact disguises the inherently exploitative and oppressive relationship that characterizes capitalism. This relationship is explained by Marx's modification of Ricardo's iron law of wages. According to Ricardo, capitalists—driven by the need to make profits and generate more capital—

44 Ellen Meiksins Wood, *Empire of Capital* (London: Verso, 2003); see also Richard Peet, *Global Capitalism: Theories of Societal Development* (New York: Routledge, 1991) and Paul D'Amato, *The Meaning of Marxism* (Chicago: Haymarket Books, 2006).

45 Wood, *Empire of Capital*, 10.

46 David Harvey, *Spaces of Capital* (New York: Routledge, 2001), 237.

47 Baradat, *Political Ideologies*, 186.

will pay their workers only subsistence wages. Anything that workers produce above the subsistence level becomes surplus value, earnings that the capitalists keep as profit. Marx reasoned, however, that since surplus value can only be produced by labor, it belongs by right to the laborer. Put differently, workers produce enough value to cover the cost of their wages (variable capital) in just a part of the working day; the labor performed for the rest of the working day does not have to be paid for—it is "surplus labor" which produces "surplus value." When the product is sold, this unpaid portion goes directly to the capitalist.[48] Any profit the capitalists make, therefore, from the labor of their employees is ill-gotten and exploitative.[49] It bears noting that Marx did not reject profit nor oppose it in principle; rather, he rejected the capitalist and he opposed private profit. Profit, according to Marx, should be distributed in a just process.

Given the basic exploitative relationship within capitalism, Marx maintained that this form of economic organization was inherently alienating. This alienation occurs in three fundamental ways. First, Marx reasoned that since work can be a form of "self-creativity," it should therefore be enjoyable. However, because the capitalists attempt to squeeze every ounce of profit from the sweat of the laboring class, the owners make conditions of work intolerable. Consequently, members of the proletariat grow to hate the very process by which they could refine their own natures. They become, in effect, alienated from a part of their own selves. Second, given that capitalists *must* exploit workers in order to produce a profit, they also must force workers to sell the product of their labor and, in turn, use that product against workers to exploit them further. Again, the workers begin to regard their own product as alien and harmful. Lastly, in an attempt to reduce labor costs (wages) and thus to garner more profit, the capitalist strives to mechanize the production process. However, the replacement of living labor with machines further de-humanizes the entire production process and robs laborers of their skills. Workers are reduced to mere cogs in a larger system and thus endure yet another form of alienation.[50]

In dialectical fashion, Marx surmised that capitalism contained its own seeds of destruction. According to Marx, capitalists invest money to make a profit; however, this requires that capitalists sell their products on the market. There must be a demand. Herein lies the problem. Individual capitalists do not in fact control the market. Aside from certain state interventions, there is generally no central control, no planning. Rather, market transactions and commodity exchanges are predicated on competition. Capitalists continually seek to reduce production costs to cheapen their products and thus outsell their competitors. The objective is not simply to sell what they produce, but to sell at a profit. Such a process sets up a never-ending quest for the accumulation of more and more capital. But how is this possible? To garner more profits, capitalists resort to lowering wages. Consequently, as a smaller portion of investment goes to wages, a bigger portion emerges as surplus value. Another tactic is to make laborers work longer or work more efficiently. Capitalists may also invest in more productive machinery. Capitalists must also ensure that workers

---

48 D'Amato, *Meaning of Marxism*, 56.
49 Baradat, *Political Ideologies*, 187.
50 Baradat, *Political Ideologies*, 185.

continuously consume products, despite receiving lower wages. This, capitalists may achieve, by offering fictitious capital (credit). Continued purchasing of goods is also ensured through the production of goods with built-in obsolescence. The end result of all of these decisions, strategies, and tactics is that capitalism produces a vast army of unemployed or underemployed workers who are barely living at a subsistence level. For Marx, this alienated proletarian class is to provide a negation of the capitalist system.[51]

In short, Marx predicted an *inevitable* transformation that would replace capitalism and the market economy. There would emerge a socialist-based economy, with a corresponding reallocation of resources and products. Furthermore, this transformation, which would replace "commodity production" (production by independent, or private, producers for exchange on the market) with socialist organization (production for the direct use of the socialist community by producers directly associated through a consciously formulated plan) was to provide a basis for a radically different relationship between people. From this transformation, a higher form of social existence would arise.[52] This is an important point that will figure prominently in later chapters. As Roberts explains, from a Marxist stand-point, once social relations between people cease to be determined by commercial principles and once people achieve conscious control over the material conditions of life, they would then achieve self-realization and end their alienated existence.[53] In other words, *the revolution would transform not only the political economy of a society, it would also necessarily transform the very essence of humanity.*

Marx argued that the revolution would arise spontaneously; that it was an historical inevitability. An economic determinist, Marx was convinced that, through a dialectical process, economic change forces social change, which in turn leads to political change. He reasoned that through increased competition the capitalist class would be forced to purchase more machinery in order to sell more commodities at lower cost. However, since only human labor produces surplus value, the profits of capitalists would steadily decline since they would employ fewer and fewer workers. Consequently, unemployment would rise and the size of the proletariat and depth of misery would increase. Inequalities between the rich and the poor would become more pronounced. As the proletariat awoke to their plight, as they developed a communal class consciousness, they would realize the depth of their oppression and rise up to overthrow the capitalist class.[54]

In theory, Marx's historical materialism predicted a proletarian revolution. Such an inevitable revolution, however, did not spontaneously emerge in Europe, or anywhere else, for that matter. To understand why not, it is necessary to look deeper in the machinations of capitalism.

As a system, capitalism is inherently and inevitably crisis-prone. This occurs because economic growth under capitalism is a process of internal contradictions.

---

51 D'Amato, *Meanings of Marxism*, 58–59; Ingersoll et al., *Philosophic Roots*, 139.

52 Paul Craig Roberts, "'War Communism': A Re-examination," *Slavic Review* 29(1970): 238–261; at 239.

53 Roberts, "War Communism," 239.

54 Baradat, *Political Ideologies*, 189.

For example, in order to accumulate capital (profit maximization), a factory owner will pay the least amount in wages to his or her employees—Ricardo's iron law of wages. However, in light of competition from other factories, workers are encouraged to produce more and more goods. In time, though, with reduced (or "survival"-level wages), workers are unable to consume commodities at a rate commensurate with the production of goods. This leads to a situation characterized by an overproduction of capital and/or an under-consumption of products.[55] These crises of over-accumulation (also referred to as *over-production* or *over-capacity*) are marked by chronic unemployment and underemployment, capital surpluses and lack of investment opportunities, falling rates of profit, and a lack of effective demand in the market.[56]

Confronted with a crisis of over-accumulation, states may intervene in any number of ways. Of particular importance is a dynamic David Harvey terms a "spatial fix." Through this process, excess capital and labor may be absorbed through geographic expansion. Capitalism is "spread" to new territories through trade, foreign investment, and the outsourcing of employment. Colonialism, for example, has historically been a prime means to expand capitalism in the search for profits. And it was from this observation—of the relentless expansion of capitalism—that Vladimir Lenin identified as a partial explanation for why the socialist revolution did not materialize. Marx, it will be recalled, predicted a proletarian revolution. However, by the turn of the twentieth-century, conditions of labor in the industrial countries were actually improving, a trend that seemingly stood Marxist theory on its head. It was Lenin who provided an answer. According to Lenin, a new form of capitalism had emerged.

The accumulation of capital is profoundly geographic.[57] However, the late nineteenth and early twentieth-century witnessed an unprecedented spatial expansion of capitalism as a political-economic system. Domestically, capitalists found it increasingly difficult to garner desired profits from their home markets. New markets, fresh caches of resources, and cheaper labor were continuously sought. It was during this period that many industrialists, bankers, and financiers pursued their accumulation strategies in the distant lands of Africa and Asia. These years constituted the high-water mark of colonial enterprise. On the heels of European explorers—Burton, Baker, Speke, Livingstone, Stanley—came agents of chartered companies: the Royal Niger Company, the German Colonization Society, the United Africa Company, the Imperial East Africa Company, and so on. The recently formed geographical societies (e.g., Paris, 1821; Berlin, 1827; London, 1831), coupled with these new commercial and colonial societies and groups of private investors, saw possibilities for profit in commerce, trade, and the control of mineral resources.[58] In

---

55 For further discussions on crises of over-accumulation, see Harvey, *Spaces of Capital*; David Harvey, *The Limits to Capital* (Oxford: Basil Blackwell, 1982); David Harvey, *The Condition of Postmodernity* (Oxford: Basil Blackwell, 1989).

56 Walden Bello, *Dilemmas of Domination: The Unmaking of the American Empire* (New York: Henry Holt and Company, 2005), 4; Harvey, *Spaces of Capital*, 240.

57 David Harvey, *Spaces of Hope* (Berkeley: University of California Press, 2000), 23.

58 Philip W. Porter and Eric S. Sheppard, *A World of Difference: Society, Nature, Development* (New York: The Guilford Press, 1998), 323.

Indochina, for example, the privately endowed Paris Geographical Society sponsored the exploration of the Mekong under the leadership of two naval officers, Francis Garnier and Doudart de Lagrée. It was this expedition that facilitated France's eventual consolidation and control of Cambodia as part of the Franco-Khmer treaty of 1863.[59]

Lenin coined the phrase *imperialist capitalism* to describe this newest phase of capitalism and it was this re-formed system that worked to delay the proletarian revolution. This occurred because capitalists were, ultimately, able to "buy" off their domestic workers while still garnering substantial profits. Geographic expansion was the key. Capitalists, motivated to increase profits, were hindered by diminishing home markets and an increasingly agitated proletariat. Indeed, there is evidence suggesting that many capitalists took Marx's predictions seriously: as work and living conditions deteriorated, the threat of rebellion increased. Consequently, capitalists were able to diffuse the threat of labor unrest, and simultaneously expand their markets, through colonial projects. These overseas missions served to reduce the tensions produced by their previous domestic exploitation as capitalists shared *some* of their profits with their domestic workers. In effect, some of the profits reaped from their colonies were re-directed to the workers in the home country. This accounted for the marginal improvement in workers' lives exhibited in the industrialized states. Domestic workers—fueled by nationalism and racism—became partners in the capitalist exploitation of the colonized people. It was a system designed for robbing Peter to pay Paul.[60]

Imperialist capitalism provided a temporary spatial fix to capital's present crises. It was not a panacea. Indeed, Lenin surmised that imperialism was the final stage of capitalism, arguing (in 1916) that in time, all territories would ultimately be subdued and that the capitalist states would turn on each other. The First World War, from Lenin's perspective, was a monumental struggle between capitalists that would, in the end, result in an international socialist victory.[61] Lenin, though, was only partially correct. In Russia a socialist revolution did in fact materialize, resulting in the birth of the Union of Soviet Socialist Republics (USSR). Elsewhere, though, capitalist governments were more resilient. The victorious states of England, France, and the United States, following their victory over Germany in the First World War, continued their colonial practices throughout Asia and Africa. In Cambodia, for example, French colonial authorities tightened and rationalized their control over the country. Taxes remained extraordinarily oppressive, a legacy of policies enacted during the First World War, when the French increased the tax burden throughout

---

59 D.R. SarDesai, *Vietnam: Past and Present*, 4th edition (Cambridge, MA: Westview Press, 2005), 38. SarDesai, who describes the society as a "front for business interests," notes also that the president of the Paris Geographical Society at the time was Chaseloup, head of the Ministry of the Navy. Furthermore, in 1873, in cooperation with the Paris Chamber of Commerce, a Society for Commercial Geography was established that sponsored and financed overseas missions. Later, Garnier, along with Jean Dupuis, were able to facilitate the French conquest of the Red River in northern Vietnam and declared three Vietnamese ports open to foreign commerce. A French consul and garrison was stationed in each (pp. 39–41).

60 Baradat, *Political Ideologies*, 200.

61 Baradat, *Political Ideologies*, 200.

Indochina to pay for the conflict. As Chandler elaborates, the French financed almost all of their activities in Cambodia, including public works and the salaries of French officials, by a complex and onerous network of taxes on salt, alcohol, opium, rice and other crops, and exported and imported goods and by levying extensive fees for all government services. These remained throughout the inter-war period (1918–1939), with the gap between the French and the Cambodians widening.[62]

As detailed in Chapter 2, deteriorating living conditions in Cambodia under French colonial rule led to sporadic peasant uprisings. None of these, however, can be considered an example of a socialist revolution. Cambodia—according to Marxist reasoning—was far from primed for revolution. Unlike neighboring Vietnam, there was not a sufficient laboring class that was "conscious" of its exploitation. To be sure, many Khmer resented the presence of French administrators and troops in their country. Missing was a cadre of individuals who would serve as vanguard for a truly radical resistance to French authority.

Cambodia, it will also be recalled, was politically quiescent throughout the first four decades of the twentieth-century. It was not until the late 1940s, spurred by their Vietnamese compatriots, that a small number of Khmer began to agitate for revolution. Nationalists such as Son Ngoc Than, Son Ngoc Minh, Dap Chhuon served—through action, if not in name—as vanguard for their country. Their activities, and in particular their resistance to French rule, conforms with Lenin's understanding of revolution. It was Lenin, for example, who argued that the proletariat would not develop class consciousness without the intervention of a revolutionary group. He envisioned this "vanguard of the proletariat" as the principal revolutionary agent that would overthrow the government. Furthermore, in contradistinction to Marx, Lenin surmised that the vanguard could actually trigger a revolution long before optimal conditions (as theorized by Marx) existed. It was this reasoning that fueled the socialist revolutions in both Vietnam and China and, in turn, led to the formation of both the Khmer People's Revolutionary Party (KPRP) in 1951 and the Saloth Sar-dominated Communist Party of Kampuchea (CPK).

Manufacturing the Revolution

The Cambodian revolution, as envisioned by Saloth Sar and his associates, provides a text-book lesson of praxis. It was Marx who argued that "The Philosophers have only interpreted the world, in various ways; the point is to change it." Marx continued that:

---

62 David Chandler, *A History of Cambodia*, 3rd edition (Boulder, CO: Westview Press, 2000), 153–155. Chandler notes, for example, that a French official could earn as much as 12,000 piastres a year, and with exemptions for his wife and two children, would pay only 30 piastres in tax. A Cambodian farmer, conversely, might earn only 90 piastres a year, and yet was required to pay 12 piastres per year in taxes. His rice, moreover, was taxed at a fixed percentage, regardless of the yield. Droughts and floods could (and did) result in widespread famine.

The materialist doctrine concerning the changing of circumstances and upbringing forgets that circumstances are changed by men and that it is essential to educate the educator himself. This doctrine must, therefore, divide society into two parts, one of which is superior to society. The coincidence of the changing of circumstances and of human activity or self-changing can be conceived and rationally understood only as revolutionary practice.[63]

As the above writings indicate, Marx was highly critical of all thinkers who constructed abstract idea systems that seemed to have no direct connection to material reality.[64] What was important was to initiate change, hence the necessity of a vanguard. For Marx, the vanguard was to assume a secondary role, one that served to instill an understanding of their exploitation and oppression. He did not advocate that the vanguard should organize and lead the revolution. On this point, Lenin, as well as Joseph Stalin, Leon Trotsky, and Mao Zedong, differed considerably. For these men, a leading vanguard was essential.

Lenin actively reformulated Marxist theory to conform with the actualities of Russia. Given the lack of a large proletariat, and the observation that the proletariat—upon assuming even limited political power—tended to adopt bourgeois values, Lenin argued for the necessity of a small, tightly organized band of "professional" revolutionaries to lead the movement. Lenin once explained that "no revolutionary movement can endure without a stable organization of leaders maintaining continuity" and that "the broader the popular mass drawn spontaneously into the struggle, which forms the basis of the movement and participates in it, the more urgent the need for such an organization."[65]

Within Cambodia, this lesson was widely understood, and followed, by both those Khmer leaders associated with the Vietnamese communists, as well as the Paris-educated Khmer. The latter, in particular, worked tirelessly in their effort to lay the groundwork for the revolution. Throughout the late 1950s many Khmer communist members were pursing teaching careers throughout Cambodia. Saloth Sar, for example, taught French, history, geography, and civics at the newly established and privately-run collège, Chamraoun Vichea ("Progressive Knowledge"), in Phnom Penh, while his wife, Khieu Ponnary taught Cambodian literature at Lycèe Sisowath. Son Sen and his wife Yun Yat also taught at Lycèe Sisowath. Son Sen was later assigned to the US-funded Kompong Kantout Teachers' College. Thiounn Mumm became Dean of the country's Medical Faculty, while Keng Vannsak assumed the position of Dean of the Literature Faculty at Phnom Penh University. Another prominent Leftist, Hou Youn, became director of Kambuboth (Kambuj'bot, or "Son

---

63 Karl Marx, "Theses on Feuerbach," in *Karl Marx: Selected Writings*, 2nd edition, David McLellan (ed.) (Oxford: Oxford University Press, 2000), 172–173.

64 Ingersoll et al., *Philosophic Roots*, 165.

65 Ingersoll et al., *Philosophic Roots*, 152. The professional revolutionaries in Cambodia also draw inspiration from the idea of permanent revolution. Initially developed by Trotsky, and expanded by Mao, a theory of permanent revolution maintained that an on-going revolution was required even after the socialist revolution had succeeded. The Khmer Rouge, in particular, would forcefully—and brutally—hammer this idea home after their victory in 1975.

of Kampuchea"), a private school which employed a number of other Leftist teachers, including Ieng Sary. During his stay at Kambuboth, Ieng Sary taught history and geography.[66]

Aside from these formal positions, many Khmer Leftists participated in informal seminars and delivered many public speeches. Saloth Sar, while continuing his teaching duties at Chamraon Vichea, conducted numerous semi-clandestine seminars on civic virtue, justice, and corruption. Many of these "classes" were attended by monks, students, military officials, and bureaucrats. Throughout his speeches, Saloth Sar spoke of a "new society," although the Marxist-Leninist foundation of his arguments was often downplayed or concealed. Nor did Saloth Sar explicitly ask the participants to join any movement—especially the Workers' Party—but instead attempted to raise people's consciousness.[67] Rhetorically, Saloth Sar was prudential in approach; he encouraged his audience to question, evaluate, and monitor their own actions and relationships. Although such an approach was forbidden *after* the revolution, during the 1950s and early 1960s this tactic was most appropriate. Cambodian society, as the work of Ben Kiernan, David Chandler, and others has demonstrated, was far from optimal for any revolution, at least according to theory. It was incumbent upon the vanguard, from Saloth Sar's perspective, to awaken the Khmer from their colonial—and monarchical—induced slumber.

Marx vacillated over whether violence was necessary to achieve socialist goals.[68] Other revolutionaries were not so circumspect. It was Mao, for example, who argued that "a revolution is not a dinner party, or writing an essay, or painting a picture, or doing embroidery; it cannot be so refined, so leisurely and gentle, so temperate, kind, courteous, restrained and magnanimous. A revolution is an insurrection, an act of violence by which one class overthrows another."[69] Paul D'Amato explains that "Marxists understand that war is built into the very fabric of capitalism, and that it can only be abolished when the weapons of the world's ruling classes are wrested from their hands."[70] Such reasoning justified, for revolutionaries such as Lenin, Stalin, and Mao, the necessity of violence and armed struggle to accomplish the revolution. In Cambodia, members of the various communist associations pursued several strategies. Party members worked within the Democrat Party; others participated within the Pracheachon; and still others attempted to foment change from within, by serving in Sihanouk's own *Sangkum*. However, for Saloth Sar and other high-ranking members of the Khmer Rouge, the dismal results of these alternative strategies demonstrated that peaceful change was futile. Neither the French nor the Americans, Sihanouk nor Lon Nol, would willingly relinquish their authority; it would be next-to-impossible to bring about change through normal political channels. Only through revolution—and specifically armed warfare—would change occur. For Marxists such as Saloth Sar, therefore, it is important to recognize that war is progressive and justified.

---

66 Chandler, *Brother Number One*, 50–51.
67 Chandler, *Brother Number One*, 62.
68 Baradat, *Political Ideologies*, 188.
69 Quoted in Goodwin, *No Other Way Out*, 275.
70 D'Amato, *Meaning of Marxism*, 156.

Trotsky, writing in 1919, defended his use of violence. He questioned whether or not, "When a murderer raises his knife over a child, may one kill the murderer to save the child? Will not thereby the principle of the "sacredness of human life" be infringed? … Is an insurrection of oppressed slaves against their masters permissible?" Trotsky countered that "As long as human labor power and, consequently, life itself remain articles of sale and purchase, of exploitation and robbery, the principle of the "sacredness of human life" remains a shameful lie, uttered with the object of keeping the oppressed slaves in their chains … To make the individual sacred we must destroy the social order which crucifies him."[71] What Trotsky provides is a defense of killing; he argues that the taking of an individual life is justified if it liberates an oppressed and exploited class. In short, it sanctifies and justifies *some* lives at the expense of others. As detailed in Chapter 6, this reasoning provided the rationale that led to the death of over two million people in Cambodia.

A justification of violence was not the only insight derived from the re-workings of Marxist theory. The vanguard of the Khmer Rouge—Saloth Sar, Ieng Sary, Nuon Chea, Son Sen, and Khieu Samphan, in particular—recognized also that Marxist-Leninist theory must be adapted to fit the particularities of the situation. Accordingly, the Khmer Rouge borrowed heavily (despite their avowals to the contrary) from Mao and the Chinese revolution. Unlike Russia, which exhibited at the time of its revolution a small, yet important, level of industrialization, China was decidedly agrarian. Any revolution, therefore, would be predicated on the mass mobilization of the peasantry. Mao, therefore, expanded upon Lenin's contention that a revolution is possible—through the efforts of a vanguard—despite the fact that capitalism had not progressed to a point where a sizeable proletariat had formed. For Mao, it became imperative to "proletarianize" the peasantry. He believed that the peasants' simple and pure character, untarnished by capitalism, would provide the necessary strength. Correspondingly, Mao privileged ideological purity over economic training; it was his contention that a proletarian mentality could be developed through educational and well as economic experience.[72]

In theory, the Khmer Rouge exemplified this idolatry of the peasantry; in actuality, the practices of the Khmer Rouge led to an intense suffering of the masses. In word, but not in deed, the Khmer Rouge attempted to mold Cambodia's citizens into perfect specimens of ideological purity. This was to be accompanied by another modification of Marxism, that being the relationship between nationalism and communism. As exemplified in the writings of Lenin, Stalin, Trotsky, and Mao, nationalism and communism sit uneasily. On the one hand, nationalist sentiment proves highly effective in mobilizing a population. This is seen especially in places like Vietnam and Cambodia, where nationalism was promoted as part of an anti-colonial movement. On the other hand, a promotion of national identity runs counter to the Marxist ideal of an international socialist revolution. Consequently, different revolutionaries leaders would support, resist, or tolerate a fusion of nationalism and communism, depending on the particular local conditions at hand.

---

71 Quoted in D'Amato, *Meaning of Marxism*, 157
72 Baradat, *Political Ideologies*, 214.

Within Cambodia, as discussed in Chapter 3, the Khmer Rouge played up Khmer identity for two main reasons. First, it was far more effective to promote the revolution as anti-colonial as opposed to pro-communist. Second, Saloth Sar and the other Paris-educated communists were fearful of their Vietnamese and Vietnamese-influenced Khmer communist cadres. The Khmer Rouge were resentful of the Vietnamese Communist Party's attempt to suppress the Cambodian revolution. Significantly, these two factors worked together, given that many Khmer peasants were already prepared to believe that Vietnam harbored plans to colonize Cambodia.

In short, the Paris-educated Khmer embarked on a revolution that was theoretically diffuse, drawing elements from Marx, Lenin, Trotsky, and Mao, among others. They promoted themselves, and operated as, a small cadre of professional revolutionaries; they sought to create the conditions necessary for revolution through a mobilization of the peasantry; they emphasized national identity—and downplayed international solidarity—both as a means of retaining exclusive control of the movement as well as to mobilize the peasantry; they advocated permanent revolution; and they believed in the necessity of armed struggle.

The Khmer Rouge operated on the belief that there could never be a true peace or a permanent accommodation with capitalism because the two systems diametrically contradict each other.[73] The lesson was clear; the Khmer Rouge were required to erase all vestiges of the previous economic system, including both the base (e.g., the means and relations of production) as well as the superstructure (e.g., arts, literature, religion, governance, and so forth).

After the Revolution: The Spaces of Ground Zero

In considering the geographical imaginations of the Khmer Rouge, I am struck by Stephen Louw's prescient question: What happens if a party—committed to the abolition of commodity production and the introduction of communism—comes to power *when the conditions supportive of a communist revolution are not present and did not cause the party's seizure of power?*[74]

That the Khmer Rouge maintained their existence, let alone achieved victory, is truly remarkable.[75] When the Khmer Rouge stood victorious on the streets of Phnom Penh, on April 17, 1975, they constituted neither a centralized, efficient political party nor military force. Their victory was the haphazard by-product of the culmination of a series of concurrent revolutions and social movements: anti-colonial, anti-monarchical, anti-American, and anti-Vietnamese. Through their own brutal suppression of dissenters and opposition, the Khmer Rouge assumed control of a war-torn and exhausted country. They defeated the French, the Americans, and

---

73 Baradat, *Political Ideologies*, 215.

74 Stephen Louw, "In the Shadows of the Pharaohs: The Militarization of Labour Debate and Classical Marxist Theory," *Economy and Society* 29(2000): 239–263; at 240.

75 Such an observation should not be construed as an endorsement or legitimation of the Khmer Rouge. Rather, this observation highlights how poorly prepared and out-of-touch the Khmer Rouge were with Cambodian society. These deficiencies would, tragically, manifest themselves in policies and practices that resulted in the death of millions.

the Khmer Republican forces not because of their own military superiority (despite their claims to the contrary) but because of military blunders by Lon Nol and of the military prowess (initially) of the Vietnamese communist forces. The Khmer Rouge "achieved" victory not because they were united in principle and in ideology with the Khmer populace; in fact, the socialist revolution enacted by the Khmer Rouge was not the end-result of a popular uprising. It was not, in this sense, a "pure" communist revolution. Instead, it was the result of a repressive cadre that capitalized on the death and destruction that engulfed Cambodia.

This returns us to Louw's question: What happens if a party comes to power when the conditions supportive of a communist revolution are not present? Louw suggests that revolutionaries are faced with two choices. On the one hand, they may choose to abandon communist doctrine and explore alternative strategies or programs. On the other hand, they can remain faithful to the central tenets of classical Marxism and thus attempt to introduce communism to the populace through administrative fiat. In other words, fundamental economic, social, and political transformations would be introduced into society, by force if necessary.[76]

The Khmer leadership, based on their eclectic understanding of Marxist-Leninist doctrine, understood that they would have to "build socialism." This concept, Thion explains, is derived from the theorizing of early post-Leninist Marxian economists who realized that revolutions do not always take place after the final phase of capitalist evolution and, therefore, the concentration of productive forces must be achieved under revolutionary control. Consequently, according to these economists, in order to achieve socialism and proceed toward a communist utopia, transition phases must be identified and implemented. Thion argues, however, that the Communist Party of Kampuchea (CPK) grossly misunderstood the idea of "transitional phases" as "reformist" or "revisionist." Such a strategy, from their vantage point, was to be avoided. Instead, the Khmer leadership believed that they would be able to jump over the transition phases—to bring about a *super great leap forward*—and achieve instant socialism.[77] The transformation of Cambodia was, from the perspective of the Khmer Rouge leadership, to literally crush all pre-existing histories, geographies, and societies. Before a utopian communist society could be constructed, the spaces of Cambodia required an un-making.

"Space" is ubiquitous in both the "social" and "physical" sciences. As Mike Crang and Nigel Thrift explain, "Space is the everywhere of modern thought." They caution, however, that the problem of space is not so much that space means very different things, but that it is used with such abandon that its meanings run into each other before they have been properly interrogated.[78] A central argument I forward in *The Killing of Cambodia* is that the Khmer Rouge leadership demanded an *un-*

---

76 Louw, "Shadows of the Pharaohs," 240.

77 Serge Thion, "The Cambodian Idea of Revolution," in *Revolution and its Aftermath in Kampuchea: Eight Essays*, David P. Chandler and Ben Kiernan (eds) (New Haven, CT: Yale University Southeast Asia Studies, 1983), 10–33; at 25.

78 Mike Crang and Nigel Thrift, "Introduction," in *Thinking Space*, Mike Crang and Nigel Thrift (eds) (New York: Routledge, 2000), 1–30; at 1.

making of space as a means of ushering in a communist society. This requires some clarification and specification on the meaning of space.

Since Geography's inception as a discipline in the early twentieth-century within the Anglo-American university setting, space has often been treated in absolute terms.[79] In part, this is explained by—and accounts for—the dominance of *regionalism* within Geography. Prior to the 1930s, most geographers did not seriously contemplate the "nature" of space, nor of Geography for that matter. In 1939, however, the publication of Richard Hartshorne's widely popular book *The Nature of Geography* defined for a generation the basis of Geography as a discipline and also how space was to be conceived. Drawing on the earlier writings of Immanuel Kant (1724–1804), and Alfred Hettner (1859–1941), Hartshorne defined geography as *chorology*, and it was Geography's purpose to study the causal relationships between objects occurring within particular *regions*.[80] However, as Olaf Kuhlke recognizes, for Hartshorne it was "the phenomena themselves—not the logic behind their spatial arrangement—[that were] at the center of the study. Emphasis was placed on the uniqueness of spaces and regions; conceptually, therefore, space was based on absolute fixed entities, on the arrangement of objects anchored in space in a scientific manner."[81]

Throughout the 1940s and 1950s the Hartshornian conception of space was gradually, and fundamentally, transformed. Spurred by the challenge initiated by Fred Schaefer, the focus on the uniqueness of phenomena distributed across space was re-directed. Increasingly, geographers called for an attention to the "spaces" between objects. For Schaefer, geographers needed to study the spatial arrangement of the phenomena; spatial relations were of the utmost importance, rather than the objects per se. Such an attitude crystallized in what became known as the "quantitative revolution." A new generation of geographers, located primarily at the Universities of Washingon, Iowa, and Wisconsin (Madison), began to elaborate a series of core geographic themes that were based on *relative* concepts of space. Kulke explains that these geographers, emphasizing the replicability of research and methodological clarity, outlined distance, pattern, position, and site (among others) as the basic concepts of geography.[82] Driven by the assumption that there are

---

79 For more in-depth discussions on "space" within the discipline of Geography, see Crang and Thrift, *Thinking Space*; Yi-Fu Tuan, *Space and Place: The Perspective of Experience* (Minneapolis: University of Minnesota Press, 1977); Robert D. Sack, *Conceptions of Space in Social Thought: A Geographic Perspective* (Minneapolis: University of Minnesota Press, 1980); Phil Hubbard, Rob Kitchin, and Gill Valentine (eds), *Key Thinkers on Space and Place* (Thousand Oaks, CA: Sage Publications, 2004); and John Agnew, "Space: Place," in *Spaces of Geographical Thought: Deconstructing Human Geography's Binaries*, Paul Cloke and Ron Johnston (eds) (Thousand Oaks, CA: Sage Publications, 2005), 81–96.

80 The term is derived from *choros*, meaning "place," and *graphos*, meaning "to draw, or describe." Chorology, therefore, was the scientific study of the relationship between phenomena that occur in the same place. Chronology, which defines History as a Discipline, is the scientific study of the relationship between events that follow each other in time.

81 Olaf Kuhlke, "Human Geography and Space," in *Encyclopedia of Human Geography*, Barney Warf (ed.) (Thousand Oaks, CA: Sage Publications, 2006), 441–444; at 442.

82 Kuhlke, "Human Geography," 442.

preexisting physical laws that can be scientifically measured, space was effectively reduced to the essence of geometry. Simply put, "in geographic terms, absolute space [was] defined and understood through Euclidean geometry (with x, y, and z dimensions) and, for analytical purposes, treated as an objective, empirical space."[83] Rob Shields concurs, noting for example, that "the realm of the spatial has often been assumed to be purely neutral and a-political, conferring neither disadvantage, nor benefit to any group. This "empirical space" is complacently understood to be fully defined by dimensional measurements (height, width and breadth) and by trigonometric descriptions of the geometrical relationships between objects, which are thought to sit in a kind of vacuum."[84]

Although this abstract conception of space remains dominant in many geographic centers of learning, as well as many sub-fields of the discipline, it has not been immune from challenges. Throughout the 1970s, but especially into the 1980s, a number of criticisms have emerged, most of which have sought to replace a *relative* understanding of space with a *relational* understanding. Rather than conceiving space as an inert backdrop, a stage on which humans (and other objects) operate according to abstract physical laws, space is now increasingly understood as an actor in its own right. Space, in effect, is *produced*; but so too does space *produce*. Further elaboration is in order.

It is common-place, for example, to say that an individual—such as Saloth Sar—made history. Implicit in this statement is the understanding that "history" was made discursively; that through one's actions, events were affected. Less common, though, is the assertion that an individual "makes space." Such a notion, however, lies at the core of a relational understanding of space. As such, Doreen Massey asserts that "Space is constituted through social relations and material social practices."[85] This is captured also in David Delaney's idea of *geographies of experience*. Delaney writes, "Our lives are, in a sense, made of time. But we are also physical, corporeal, mobile beings. We inhabit a material, spatial world. We move through it. We change it. It changes us. Each of us is weaving a singular path through the world. The paths that we make, the conditions under which we make them, and the experiences that those paths open up or close off are part of what makes us who we are."[86]

Delaney prefigures a discussion of a dialectics of self and space—a subject that will reappear in later chapters. For now, it is important to acknowledge that through our daily activities we encounter other peoples and other places; our thoughts and actions are influenced by these encounters. Concurrently, however, our presence, our interactions, likewise reflect back upon those spaces. Just as we are transformed

---

83 Phil Hubbard, Rob Kitchin, Brendan Bartley, and Duncan Feller, *Thinking Geographically: Space, Theory and Contemporary Human Geography* (London: Continuum, 2002), 13).

84 Rob Shields, "Spatial Stress and Resistance: Social Meanings of Spatialization," in *Space and Social Theory: Interpreting Modernity and Postmodernity*, Georges Benko and Ulf Strohmayer (eds) (Malden, MA: Blackwell, 1997), 186–202; at 187.

85 Doreen Massey, *Space, Place, and Gender* (Minneapolis: University of Minnesota Press, 1994), 254.

86 David Delaney, *Race, Place and the Law, 1836–1948* (Austin: University of Texas Press, 1998), 4.

through our daily activities, so too are the spaces which we inhabit transformed. In short, we produce and are produced by space.[87] However, as Henri Lefebvre cautions, space is not produced in the sense that a kilogram of sugar or a yard of cloth is produced. Rather, space is "the product of competing ideas (discourses) about what constitutes that space order and control or free, and perhaps dangerous interaction."[88]

A relational conception of space directs attention to how space is constituted and given meaning through human interactions. To this end, Ed Soja has introduced the term "spatiality" in reference to socially-produced space. Soja explains:

> The dominance of a physicalist [i.e., abstract or relative] view of space has so permeated the analysis of human spatiality that it tends to distort our vocabulary. Thus, while such adjectives as 'social,' 'political,' 'economic,' and even 'historical' generally suggest ... a link to human action and motivation, the term 'spatial' typically evokes a physical or geometrical image, something external to the social context and to social action, a part of the 'environment,' a part of the setting for society—its naively given container—rather than a formative structure created by society. We really do not have a widely used and accepted expression in English to convey the inherently social quality of organized space, especially since the terms 'social space' and 'human geography' have become so murky with multiple and often incompatible meanings.[89]

Shields follows with a further justification for a conception of space as relational. "If one still bridles," he argues, "at the idea of a social 'production' or cultural 'making' of 'spaces' then perhaps one might refer to the remaking of empirical space by social groups." This remaking of space, he explains, "takes place almost invisibly" because "the social categories in which space is conceived and perceived structure the most elementary aspects of our interaction with our physical context and setting." Consequently, the "juggling act of 'making space' and putting into practice spatial codes is indicative of a larger social quality to spatial coding, to spatial practices, to our representations of this 'space,' and to our 'imaginary geography' in which everything has a place and a time."[90]

Arguably, it has been the theoretical insights of Henri Lefebvre that have most revolutionized the conception of space. According to Shields, Lefebvre sought an approach to understand the dialetical interaction between spatial arrangements and social organization through a refusal to theorize space in terms of its own codes and logic. For Shields, therefore, it is essential to acknowledge that much of Lefebvre's discussion is framed within an explanation of the production of space: how can

---

87 James A. Tyner, *The Geography of Malcolm X: Black Radicalism and the Remaking of American Space* (New York: Routledge, 2006), 63.

88 Don Mitchell, "The End of Public Space? People's Park, Definitions of the Public, and Democracy," *Annals of the Association of American Geographers* 85 (1995): 108–33; at 115.

89 Edward W. Soja, *Postmodern Geographies: The Reassertion of Space in Critical Social Theory* (New York: Verso, 1989), 80.

90 Shields, "Spatial Stress and Resistance," 186–187.

spatialization can be both a product and a productive medium in which other products are created and in which exchanges take place.[91]

Lefebvre proposed that every society—and hence every mode of production with its sub-variants—produces a space, its own space. Transforming Marx's periodization of capitalism into a history of space, Lefebvre showed how different relations between those elements produced different "spaces."[92] He argued that every system of economic organization, whether feudalism or slavery, merchant capitalism or industrial capitalism, will be manifest on the landscape. However, space is neither an "*a priori* condition" of institutions or structures, nor it is "an aggregate of the places of locations" of phenomena or products. Space, according to Lefebvre, is *produced* via competing discursive claims, usages, and material practices.[93] Significantly, Lefebvre framed his arguments within the context of contemporary capitalism; he attempted to provide a spatiality to Marx's dialectic materialism. Lefebvre accomplished this through the forwarding of a "three-fold dialectic" within spatialization,[94] which included *spatial practices, representations of space*, and *spaces of representation*.

Spatial practice refers to "the production and reproduction of specific places and spatial 'ensembles' appropriate to the social formation." Shields elaborates that spatial practices "include building typology, urban morphology and the creation of zones and regions for specific purposes: a specific range of types of park for recreation; test sites for nuclear weapons; places for this and that; sites for death (graveyards) and remembrance (memorials, battlegrounds, museums, historic walks and tours). Through everyday practice, 'space' is dialectically produced as 'human space.'"[95] Representations of space are conceptualized spaces, the spaces of scientists and planners, the "dominant" space in society.[96] Space, consequently, is "purposefully representational of certain societal ideas, and therefore the holders of these ideals attempt to control its use."[97] We are socialized, for example, into an understanding of these representations of space, of whom is permitted access, and what behaviors are acceptable. Public parks, we are taught, are *children's* or *families'* spaces. Teenagers who attempt to appropriate these spaces for their own use, such as for "hanging out," are shunted away by authorities. These divisions of space, and of appropriate use, become naturalized and normalized. They are also coded by dominant conceptions of "race," sex, gender and so forth. Consequently, these spaces remain highly regulated and policed; spaces become sanctioned. Representations of space, furthermore, may be materially demarcated, as in the erection of walls, signs, and fences. Enforcement may be further ensured through collective action and the threat, if not actual use, of

---

91 Rob Shields, *Lefebvre, Love & Struggle: Spatial Dialectics* (New York: Routledge, 1999), 157–159.

92 Hubbard et al., *Thinking Geographically*, 15.

93 Lefebvre, *Production of Space*, 85.

94 Shields, *Lefebvre, Love & Struggle*, 160.

95 Shields, *Lefebvre, Love & Struggle*, 162.

96 Lefebvre, *Production of Space*, 38–39.

97 Mona Domosh, "Those 'Gorgeous Incongruities': Polite Politics and Public Space on the Streets of Nineteenth-Century New York City," *Annals of the Association of American Geographers* 88 (1998): 209–26; at 10.

force. Over time we internalize these lessons, and learn appropriate behavior. We are produced by these spaces just as we produced these spaces.

The dominance of representations of space is far from complete; these authorial representations do not go unchallenged. Spaces of representations (termed *representational spaces* by Lefebvre) are "sites of resistance, and of counter-discourses which have not been grasped by apparatuses of power."[98] According to Lefebvre, representational spaces "need obey no rules of consistency or cohesiveness. Redolent with imaginary and symbolic elements, they have their source in history—in the history of a people as well as in the history of each individual belonging to that people."[99]

These aspects operate at all times. Shields elaborates that spatializations "find their grounding in a process of production through practice governed by the influence of historico-socially constructed spaces of representation. Spatialisations are then refined and rationalised in representation and discourses on space, which act upon production practices by specifying the appropriate movements of bodies (gestures, etiquette), materials (commodities), and relations (communications) in space."[100] Consequently, according to Lefebvre, it is possible to reconfigure the specific ways in which space has been produced within particular economic (and political) organizations.

Lefebvre's periodization and his spatialization of Marx's history of economic organization has been challenged. Shields, in particular, finds that Lefebvre "overstates" his case when he attempts to delimit spatial essence of each historical epoch. Shields likewise finds fault with Lefebvre's "stereotypical, linear, Eurocentric modelling of historical progress." However, as Shields also notes, Lefebvre's real object of study was "the *process* of the production of space, and its configuration in any given historical period."[101] And it is in the process, I maintain, that Lefebvre's spatialization is most germane to a study of the Khmer Rouge's geographical imagination. In *The Killing of Cambodia* I invert Lefebvre's spatial contribution. I argue that before the Khmer Rouge *constructed* their own communist spaces, they deliberately set out to *deconstruct*, or unmake, previous spaces. This coincides with other elements of the Khmer Rouge's Democratic Kampuchea. Consider, for example, the idea of history. On April 17, 1975, after capturing Phnom Penh, the Khmer Rouge announced that date as the beginning of history. It was to be Year Zero. From that point on, a new Khmer history—freed from capitalism and colonialism, religion and the monarchy, westernization and intellectualism—was to be written. However, there existed a spatial counter-part to Year Zero, indeed, a "Ground Zero." For the Khmer Rouge leadership, it was not only necessary to write history anew, it was also necessary to write geography anew.

---

98 Lynn Stewart, "Bodies, Visions, and Spatial Politics: A Review Essay on Henri Lefebvre's *The Production of Space*," *Environment and Planning D: Society and Space* 13(1995): 609–618; at 611.

99 Lefebvre, *Production of Space*, 41.

100 Shields, *Lefebvre, Love & Struggle*, 167.

101 Shields, *Lefebvre, Love & Struggle*, 167–170.

Making the Super Great Leap

Lefebvre postulated that every "society to which history gave rise within a framework of a particular mode of production, and which bore the stamp of that mode of production's inherent characteristics, shaped its own space." As such, dominant economic systems, such as feudalism or merchant capitalism, will be manifest on the landscape. However, since "each mode of production has its own particular space, the shift from one mode of production to another must entail the production of a new space." Landscapes of feudalism give way to landscapes of merchant capitalism; landscapes of merchant capitalism are refashioned into landscapes of industrial capitalism. In an ongoing process of dialectical materialism, as revolutions transform one economic system into another, the landscape reflects the accumulated burden of previous revolutions. Consequently, an examination of the transitions between changing modes of production will reveal that new spaces are produced, spaces which are "planned and organized" in conformity with the emergent system.[102]

Lefebvre theorized also that as "social space is produced and reproduced in connection with the forces of production (and with the relations of production)," these forces, as they develop, do not take over "pre-existing, empty, or neutral space ... "[103] Rather, in a dialectic process, new forms of socio-economic organization are secreted onto the remnants of earlier forms. For the Khmer Rouge, however, in accordance with their understanding of total revolution, it was not acceptable to simply build on earlier foundations. For the Khmer Rouge, the planned and organized spaces of Democratic Kampuchea were not to be tainted by any association with Cambodia's pre-existing spaces.

In 1975 the material landscape of Cambodia would have revealed vestiges of its past histories and geographies. It would have reflected an "indigenous" pre-colonial Khmer society; but also present, unevenly, would be the trappings of French colonialism, and, to a lesser extent, an American presence. Such sequent occupances would have been most pronounced in the main cities, including Phnom Penh. But other, more peripheral regions, would have also revealed these elements. The Khmer Rouge, though, were not content with retaining past inscriptions of previous modes of production and spatial practices. Instead, the KR explicitly attempted to erase time and space to create (in their minds) a pure utopian communal society. This is seen most vividly, and brutally, in the Khmer Rouge's decision to evacuate Phnom Penh and all other urban areas of Cambodia.

Within hours of their entry into Phnom Penh on April 17, 1975, Khmer Rouge soldiers began the Herculean task of de-populating Phnom Penh and other major cities throughout Cambodia. The capital city, in particular, had swollen in size as hundreds of thousands of refugees fled the countryside in the face of intensive American bombings campaigns and civil war. Indeed, the evacuation was encouraged by the Khmer Rouge who forwarded the very plausible sounding explanation that American planes were about to bomb the city.

---

102 Lefebvre, *Production of Space*, 46–47; see also page 412.
103 Lefebvre, *Production of Space*, 77.

There was not, however, an impending bombing campaign planned by the United States. Such an assertion was merely a convenient lie disseminated by the CPK in support of a long-standing practice of emptying captured cities.[104] As the scholarship of Karl Jackson, François Ponchaud, and Kevin McIntyre reveals, the Khmer Rouge targeted urban affairs for a variety of reasons. First, there was the practical issue that the Khmer Rouge were in no position to feed and shelter the massive refugee population in Phnom Penh (not that they necessarily wanted to). More salient were security concerns, namely the need to preserve the fledgling revolution. As Jackson identifies, for the Khmer Rouge cities were centers of foreign domination; cities were dominated, economically at least, by large Vietnamese-Khmer and Sino-Khmer bureaucratic and commercial populations. Moreover, cities were dominated by the institutions that had opposed the revolution, including the monarchy, Lon Nol's army, the foreign embassies, and the bourgeoisie. A forcible evacuation was deemed the most efficient means to disorganize any potential opposition. In this way, an amorphous heterogeneous population could be re-organized—following purges of the most counterrevolutionary elements—into more readily controlled and productive units. Likewise, the depopulation of urban areas would provide a surplus of productive labor and thereby facilitate the active construction of the Khmer Rouge's utopian dreams.

There is another aspect to the policy that needs explicating, one that conforms with the geographical imaginations dreamt by the Khmer Rouge. Specifically, the depopulation of urban areas was essential to the Khmer Rouge's strategy of "revolution by eradication"[105] and, as such, it illustrates well the thoroughly modernist foundations of the Khmer Rouge leadership and their attempt to erase space.

---

104 Documentary evidence indicates that the decision to evacuate Phnom Penh and other major cities and towns occurred as early as February 1975. Indeed, the work of Karl Jackson and Kevin McIntyre details that such a decision conformed readily with the political economic practices advanced by Khieu Samphan in his 1959 doctoral dissertation. Specifically, the destruction of cities would, in a single stroke, deal an immediate blow to Cambodian involvement in the international economy which Khieu Samphan had identified as the single most obstructive barrier to an independent and self-reliant Democratic Kampuchea. Furthermore, the forced evacuations of April 1975 were not without precedent. As the Khmer Rouge increased their control in the early 1970s, they embarked upon smaller scaled de-population operations. Beginning as early as 1971, "liberated" zones were re-organized in cooperatives and collectives with the urban areas razed and emptied of their people. In September 1973, for example, CPK troops seized half of Kompong Cham city and took fifteen thousand townspeople into the countryside with them. And in March 1974 a combined force of CPK Northern and South-Western Zone troops, led by Pauk and Ta Mok, overran the former capital of Oudong, north of Phnom Penh. Having captured the city, the Khmer Rouge led the populace of twenty-thousand persons into the countryside, killed all school-teachers and government officials, and deliberately razed the town. See Ben Kiernan, *How Pol Pot Came to Power: A History of Communism in Kampuchea, 1930–1975* (London: Verso, 1985); see especially pages 371 and 384.

105 Karl D. Jackson, "The Ideology of Total Revolution," in *Cambodia, 1975–1978: Rendezvous with Death*, Karl D. Jackson (ed.) (Princeton: Princeton University Press, 1989), 37–78; at 56.

Democratic Kampuchea was not, contrary to the claims of some accounts, a return to a golden era of Cambodian history. The motivations of Saloth Sar, Ieng Sary, and Khieu Samphan was not to *re-create* the glories of Angkor Wat, but instead to make an entirely new, modern, productive communal society. Such a vision was articulated early on, as indicated by Khieu Samphan's dissertation. Entitled "Cambodia's Economy and Industrial Development," Khieu Samphan's dissertation—although not a template—would prove highly influential in subsequent policies of Democratic Kampuchea. In this document Khieu Samphan forwarded a series of interrelated theses. He stated that Cambodia's economy was backward, locked in a feudal and pre-capitalist mode of production. This condition, he maintained, resulted from Cambodia's unequal and dependent integration into the French colonial economy and the continuance of unfair trade relations. The Khmer Rouge leadership believed that most, if not all, of Cambodia's problems were derived from its subordinate position in an international economic system that was dominated by, among others, the United States, China, and France. Liberation for Cambodia's economy could only occur through autonomous development, namely a withdrawal from the international market. Lastly, it was imperative for Cambodia to confront the existent structural inequity between what he perceived to be a productive countryside and an unproductive city. In his doctoral work, Khieu Samphan described cities as unproductive sites, populated by individuals who contributed nothing to society but, in return, capitalized on the exploitation and oppression of the rural-based peasants. According to Khieu Samphan, and those who agreed with his interpretations, only agriculture, crafts, and small industry were productive. Tertiary activities, including commerce and banking were considered unproductive. Moreover, Khieu Samphan deplored those activities that serviced the bourgeois, including boutiques, household servants, and venders. In his dissertation, written seventeen years prior to the Khmer Rouge victory, Khieu Samphan had claimed that approximately 94 percent of workers in Phnom Penh, and 96 percent of workers in Kompong Cham City, were engaged in unproductive activities.[106]

Given such a geographical imagination, it should come as no surprise that the Khmer Rouge forwarded a decidedly anti-urban policy. Cities attracted the ire of the Khmer Rouge on two counts. On the one hand, cities were rabbit-warrens of vice, filth, corruption and disease. On the other hand, cities symbolized all that was wrong with Cambodia and its rightful place in the universe. Consequently, cities and all that they entailed, were to be smashed. The people were to be relocated and resettled throughout the countryside; they were to participate in a major collective effort to dramatically transform their lives and that of the state. The people of Cambodia were to "Strike through, smash, reduce everything to dust."[107]

The annihilation of urban areas dovetails with the Khmer Rouge's attempt to "build socialism" through by-passing any and all transitional phases. Thus, in their desire to make a "super great leap" into communism, the Khmer Rouge leadership justified their brutal practices of forced evacuations and resettlement schemes. Their

---

106 Kevin McIntyre, "Geography as Destiny: Cities, Villages and Khmer Rouge Orientalism," *Comparative Studies in Society and History* 38(1996): 730–758; at 74.

107 Locard, *Pol Pot's Little Red Book*, 229.

goal was to construct, immediately, a homogenous, egalitarian society. Jackson explains that "The Khmer Rouge sought not to turn back the pages of time to an earlier era ... but to rush forward at a dizzying pace regardless of the consequences."[108] This transformation entailed a literal wiping clean the slate that was Cambodia. In a most blunt and brutal manner, the Khmer Rouge leadership spoke of "cleaning up" the cities and towns of Cambodia. Cleaning up, of course, was merely a euphemism for purging the cities of oppositional and supposedly decadent and counter-revolutionary elements. In short, the spatial practices of the Khmer Rouge leadership reflect a desire not to *return* to a past geography, nor to build upon existing geographies. Rather, the Khmer Rouge advanced a genocidal practice that was deliberately, and literally, designed to erase—to clean up—all previous spaces of Cambodia.

The transformations of Cambodia envisioned by the Khmer Rouge were to be total, permanent, and pervasive. Conceptually, the spatial practices enacted by the Khmer Rouge conform most with Trotsky's notion of war communism. Similar to Cambodia, the Bolsheviks assumed power *before* the conditions supportive of a communist revolution were present. Consequently, Lenin and other revolutionaries enacted a series of policies designed, in theory, to cultivate those conditions necessary for a communal society.[109]

War communism emerged in Russia in response to civil war. There was, in effect, a perceived military urgency to ensure control over society and to re-establish law and order in rapidly disintegrating society. In the aftermath of the revolution, decisions were chosen deliberately, framed within the contours of Marx's account of dialectics and the transition to communism. It was understood, for example, that communism entailed the abolition of the labor market; there would be a period in which the revolutionary state would assume control over labor power, and that labor was to be conceived as a service to be rendered, rather than as a commodity to be bought and sold. However, the revolutionaries also recognized that it was not enough to simply conscript civilian labor. Instead, other means had to be found to organize and control labor in a useful (and supposedly scientific) manner.[110] On Lenin's instructions, Trotsky developed a doctrine of labor militarization, which he published in *Pravda* on December 17, 1919. Under the concept of labor militarization, Trotsky proposed to adapt the methods used to build the Red Army—which had proved so successful in the revolution—to the production process and labor control. Wherever possible, for example, "Military Labor Armies" would be put to work in industrial areas,

---

108 Jackson, "Ideology of Total Revolution," 59. See also François Ponchaud, "Social Change in the Vortex of Revolution," in *Cambodia, 1975–1978: Rendezvous with Death*, Karl D. Jackson (ed.) (Princeton: Princeton University Press, 1989), 151–177.

109 Louw, "Shadows of the Pharaohs," 244. In Russia, it is common to identify three phases of post-revolutionary practice: state capitalism (until mid-1918), war communism (until early 1921) and the new economic policy (introduced in March 1921). In brief, state capitalism entailed an extensive state control of the economy, the bulk of which remained under private ownership. Under war communism, however, the entire economy was nationalized near instantaneously, with both labor and capital markets abolished. The new economic policy, lastly, involved a gradual relaxation in the agricultural, consumer goods and labor markets within a context of overall state control of the economy.

110 Louw, "Shadows of the Pharaohs," 248.

taking control of the organization and administration of both civilian and conscripted labor. Under war communism and a militarized labor system, not only were workers and peasants expected to become soldiers, but soldiers were expected to become workers.[111]

The Khmer Rouge, likewise, encountered similar conditions following their victory. In 1975 Cambodia was broken, devastated by years of civil strife, warfare, lawlessness, and corruption. Confronted with these conditions, the Khmer leadership embarked upon their own strategy of war communism, one that entailed the militarization of all facets of society.[112] One slogan that echoed throughout the desolate landscape of Democratic Kampuchea effectively captures the CPK's attitude of war communism: "One hand for production, the other for striking the enemy."[113] According to Locard, this slogan was widely quoted, particularly among soldiers. Through its simplicity, the slogan "unleashed an unbridled repression, an imperious demand, imposed on all the people, save the Khmer Rouge themselves, to work like convict labor." The symbolism of the money printed (but never distributed) by the Khmer Rouge likewise reflects this policy. One side of the ten riel bill, for example, reveals a harvesting scene; on the other side, guerrillas fire a mortar. On the fifty-riel bill, one side shows women planting rice; on the other side, women are shooting a bazooka.[114] Combined, these images—of everyday landscapes—reveal particular representations of space. These were—for the planners of the revolution—ordinary scenes of an extraordinary time *and* space.

Under war communism, the people of Cambodia were transformed into a grand army of workers. Peasants were also soldiers; soldiers were also farmers. All worked and fought and died for the revolution. According to Saloth Sar, for example, the "army has a great responsibility of defending the country."[115] However, this task was not sufficient; nor was defense the only responsibility of the army. Saloth Sar explained that "the regional army works in two ways: it joins the people and works with them, and it works independently on state farms. In some places, regional commanders give it the role of building roads, bridges and so on ... [However] ... the army should not devote itself merely to building, because it also has a duty to defend the country. At the same time, the army should not be separated from the people."[116] A radio editorial explained that,

---

111 Louw, "Shadows of the Pharaohs," 248–249.

112 Chandler, *Tragedy of Cambodian History*, 245; Locard, *Pol Pot's Little Red Book*, 149.

113 Locard, *Pol Pot's Little Red Book*, 163. Another, similar slogan informed the people of Cambodia that "One hand grasps a hoe, the other, a rifle."

114 Locard, *Pol Pot's Little Red Book*, 163.

115 "Preliminary Explanation Before Reading the Plan, by the Party Secretary," introduced and edited by David P. Chandler, in *Pol Pot Plans the Future: Confidential Leadership Documents from Democratic Kampuchea, 1976–1977*, David P. Chandler, Ben Kiernan, and Chanthou Boua (eds) (New Haven, CT: Yale University Southeast Asia Studies, Monograph Series No. 33), 120–163; at 143.

116 "Preliminary Explanation," 143–144.

Implementing the slogan holding hoe in one hand and rifle in the other, male and female combatants and cadres of our revolutionary army have heightened the spirit of revolutionary vigilance, defended, with great effectiveness the state power of workers and peasants, the Kampuchean Revolutionary *Angkar*, territorial integrity, national sovereignty, and our Democratic Kampuchea. The general situation of our country is good.[117]

A speech delivered in 1976 captures further the Khmer Rouge's sense of "revolutionary will." Addressing an assembly of cadres in the Western Zone, the speaker—who, most likely, was Saloth Sar—explained that, "In 1970 the [Communist] Party had over 4,000 people throughout the country. Our armed forces were small too ... The [outside] world said we were weak, small, few; how could we win?" The speaker then indicated that the Party carried the struggle on its own, that they could not nor would not depend on others. The Khmer had their own party; they had their own army. And through a thorough and meticulous analysis, they had determined that the CPK could either "win quickly, in three to four or five years" or that "the war could extend for 10 to 15 or 20 years." Consequently, according to the speaker, the CPK *chose* to opt for the first possibility, because, in the speakers' words, they "were strong politically" and "the enemy was politically inferior to us, even though he had a strong army." The speaker then explained that "We organized forces, attacked, and won in a period of five years. This was because of the Party. If the Party had not been absolute, with no correct line on strategy or tactics, we would not have won like that."[118]

The lesson was clear. The Khmer Rouge had developed a "correct" strategy for the revolution, one that led to a quick and decisive victory over militarily superior forces. Indeed, so resounding was the victory, that it was accomplished more rapidly than any other revolutionary struggle—a blatant and obvious jab at the Khmer Rouge's Vietnamese neighbors. Such revolutionary will, coupled with the proper attitude and approach, would ensure a positive change in the lives of ordinary Cambodians. According to the speaker, "We want our country to advance very quickly so that our people advance ... " However, there was a reason for such an approach: "If we are not strong and do not leap forward quickly, outside enemies are just waiting to crush us ... For that reason we must strive to move fast."[119]

Lefebvre premised that "A revolution that does not produce a new space has not realized its full potential; indeed it has failed in that it has not changed life itself, but has merely changed ideological superstructures, institutions or political apparatuses." Lefebvre continued that "A social transformation, to be truly revolutionary in character must manifest a creative capacity in its effects on daily life, on language

---

117 Quoted in Locard, *Pol Pot's Little Red Book*, 164.

118 "Excerpted Report on the Leading Views of the Comrade Representing the Party Organization at a Zone Assembly," in *Pol Pot Plans the Future: Confidential Leadership Documents from Democratic Kampuchea, 1976–1977*, David P. Chandler, Ben Kiernan, and Chanthou Boua (eds) (New Haven, CT: Yale University Southeast Asia Studies, Monograph Series No. 33), 13–35; at 25. See also Ben Kiernan's introduction to this party document, pp. 9–12.

119 "Excerpted Report," 24.

and on space"[120] Conforming to the tenets of war communism, the problem for the people of Cambodia was "to launch offensives, to become more effective in order to win tactical victory." This was to be accomplished quickly—the people were to "strike continuously in all forms"—but also thoroughly—to "strike on a large scale and on a small scale."[121]

The Revolutionary Army would serve as a model for the Cambodian population to emulate. Laboring masses of people became "fighting forces" and work efforts were promoted in the language of offensive operations. Operationally, the citizenry of Cambodia were divided into work "forces" or "teams" based on age and sex. Adults, those aged between 14 and 50, were forced into mobile work brigades termed *kong chalat*. Males belonged to *kong boroh* and females to *kong neary*. The heaviest work performed by these two groups; they plowed fields, planted and transplanted and harvested rice; and dug and carried dirt for irrigation projects.[122] Those who were younger and unmarried usually traveled great distances from village to work in forests cutting timber or working on construction projects.[123]

Older members of Democratic Kampuchea—those 50 years-old and above—were grouped into work teams known as *senah chun*. Again, these were separated by sex, with males belonging to *senah chun boroh* and females belonging to *senah chun neary*. Those in the *senah chun* normally performed lighter work (i.e., sewing, gardening, collecting wood, caring for children) and usually remained in or close to the village. Some did, however, labor in rice fields. The children of Democratic Kampuchea, which included all boys and girls under the age of 14, were organized into work groups called *kong komar*, with boys and girls separated into *kong komara* and *kong khomarei*, respectively. These members had the lightest work, often watching after cows and buffalo, digging for planting, gathering firewood, and gathering cow dung for fertilizer.[124]

Although "Every work site is a fiery battlefield," the Khmer Rouge leadership identified the promotion of self-sufficiency in agriculture as the first battle.[125] Their line of reasoning was simple: "We fight in the field of agriculture because we have agricultural resources. We'll move to other fields when the agricultural battle is finished."[126] To meet these expectations, under the Khmer Rouge the ownership of production, and consumption became increasingly communal. Cooperatives

---

120 Lefebvre, *The Production of Space*, 54.

121 "Excerpted Report," 15.

122 Kalyanee Mam, "The Endurance of the Cambodian Family Under the Khmer Rouge Regime: An Oral History," in *Genocide in Cambodia and Rwanda: New Perspectives*, Susan E. Cook (ed.) (New Haven, CT: Yale Center for International and Area Studies, Genocide Studies Program Monograph Series No. 1, 2004), 127–171; at 134.

123 Mam, "Endurance of the Cambodian Family," 135.

124 Mam, "Endurance of the Cambodian Family," 134–135.

125 Locard, *Pol Pot's Little Red Book*, 227.

126 "The Party's Four-Year Plan to Build Socialism in All Fields, 1977–1980," translated by Chanthou Boua and introduced by David P. Chandler, *Pol Pot Plans the Future: Confidential Leadership Documents from Democratic Kampuchea, 1976–1977*, David P. Chandler, Ben Kiernan, and Chanthou Boua (eds) (New Haven, CT: Yale University Southeast Asia Studies, Monograph Series No. 33), 9–35; at 48.

were established along military lines to cultivate specified crops, including rice, vegetables, fruits, and non food crops such as cotton, rubber, and jute. According to Khmer Rouge policy, "the preparations for [agricultural] offensives to build up the country are like our past military offensives and not even as difficult. In building up the country the obstacles are direct: whether there is water or not, what kind of fertilizer, what kind of seed. As for the military battlefields, they involve sacrifices. Comparing thus, we see that there is nothing to worry about."[127]

The collection and distribution of produce and goods became centralized and the establishment of large-scale collectives, coupled with irrigation projects and the development of a system of (poorly-planned and constructed) reservoirs and dams were initiated, to foster higher produce yields, and to permit multiple crops. Relatedly, and consistent with their attempt to "modernize" Cambodia through socialist practice, the Khmer Rouge initiated a spatial practice to "rationalize" the arrangement of rice paddies. Formerly, rice paddies were broken into parcels of varying shapes and sizes, resultant in a pastiche of land use patterns. Under the Khmer Rouge, though, rice paddies were to be re-organized into regular, quadrangular plots—supposedly a symmetric and efficient spatial arrangement of farming practices.[128] Ranging in size from several hundred persons to several thousand, these communal forms were the site of concentrated labor activities that were meant to achieve economic self-sufficiency for the Khmer Rouge.

Military discipline and a collective will were required to overcome the fact that Cambodia was ill-prepared for a Communist revolution. It was explained—though often at the barrel of a gun—that "we need only be clear about the strategy and tactics and we will win a basic victory."[129] These tactics, which according to the Khmer Rouge justified all means, required extreme sacrifices. Based on the Marxist understanding that capitalism and socialism diametrically contradict each other, it was not possible for the two systems to co-exist spatially. It was not possible, for the Khmer Rouge leadership, to have a socialist mode of production occupying the same space as a capitalist mode of production. The latter would have to be "smashed" before the former could be built.

Conclusions

Despite the lessons provided by Marx, a Communist revolution in Cambodia was not inevitable. Indeed, as the historical development of the Communist Party in Cambodia reveals (Chapters 2 and 3), such a revolution was not even probable. However, despite the presence of optimal conditions required for a revolution— as theorized by Marx—the leaders of the Khmer Rouge *believed* that they could manufacture a revolution and, in the process, construct a new Cambodia from the ashes of the old.

---

127 Excerpted Report," 32.

128 May Ebihara, "Revolution and Reformulation in Kampuchean Village Culture," in *The Cambodian Agony*, 2nd edition, David A. Ablin and Marlowe Hood (eds) (New York: M.E. Sharpe, 1990), 16–61; at 26.

129 "Excerpted Report," 32.

The genocide in Cambodia, consequently, was not inevitable. Neither history, geography, nor culture determined the mass killings. Colburn explains that "The defining linkage in revolutions between displacing and destroying an old order and trying to rebuild something new suggests that revolutions should be conceived not as mechanical changes of political regime or as the necessary conclusions of class conflict, but instead as the ultimate moments of political choice made only indirectly at most by the populace."[130] Governmental corruption, war, and economic hardships did not *lead* to the policies and practices that resulted in the death of millions. Rather, the mass violence was a conscious decision, one developed and enacted upon by a handful of self-proclaimed revolutionaries: Saloth Sar, Ieng Sary, Nuon Chea, Son Sen and all of the other "candidates for prosecution."[131] As Colburn elaborates, "At the moment power changes hands, the givens of social existence seem suspended, and revolutionaries' political imagination suggests that the world can be made anew."[132]

Colburn argues that "Through their language, images, and daily political activity, revolutionaries strive to reconfigure society and social relations. They consciously seek to break with the past and to establish a new nation-state. In the process, they create new social relations and, often, novel ways of practicing politics."[133] From the perspective of the Khmer Rouge, the humanity of Cambodia depended on the revolution. Accordingly, any action against their self-identified opponents was justified. The brutal policies that led to such widespread death and destruction were *legitimated* and *sanctioned* through an appeal to *social justice*. For the Khmer Rouge, a pure Communist state, one that was entirely self-reliant, one that was neither dominated nor exploited by another state, required to be made anew. This entailed the complete annihilation of all non-Communist social relations and material practices. It required that Cambodia be literally wiped clean, smashed, unmade. Only when Cambodia was killed, when the country ceased to exist both literally and symbolically, could Democratic Kampuchea be built. The unmaking of space, in effect, provided the basis for those practices that culminated in the Cambodian genocide.

---

130 Colburn, *Vogue of Revolution*, 6.

131 This phrase is borrowed from Heder and Tittemore's book, *Seven Candidates for Prosecution*.

132 Colburn, *Vogue of Revolution*, 6.

133 Colburn, *Vogue of Revolution*, 12.

Chapter 5

The Placelessness of
Democratic Kampuchea

The geographer and historian of cartography J.B. Harley asserted that "A book about geographical imagery which [does] not encompass the map would be like *Hamlet* without the Prince."[1] Harley's point was that maps are discursive. Consequently, maps as discourse "cease to be understood primarily as inert records of morphological landscapes or passive reflections of the world of objects" but rather "are regarded as refracted images contributing to a dialogue in a socially constructed world."[2] Maps, in other words, reveal much more than the lines and symbols portrayed on paper. Maps are graphic and material representations of geographical imaginations.

The work of Harley, while highly influential, was not the first attempt to consider the "political" behind cartography. During the 1940s, for example, Louis Quam and Louis Thomas provided important insights into the relationship between maps and propaganda.[3] It was not until the 1970s, however, with the arrival of Judith Tyner's dissertation on "persuasive cartography," did cartographers and geographers formally begin to *critically* evaluate and analyze maps.[4] As Norman J.W. Thrower explains, the term was coined by Tyner "to include, among others, maps used by theologians to illustrate current beliefs, by advertisers to sell products, and by propagandists to mislead the enemy."[5] Tyner's work, consequently, precipitated a still ongoing debate regarding the "authority" and "truthfulness" of maps.

---

1 J.B. Harley, "Maps, Knowledge, and Power," in *The Iconography of Landscape: Essays on the Symbolic Representation, Design and Use of Past Environments*, Denis Cosgrove and Stephen Daniels (eds) (Cambridge, UK: Cambridge University Press, 1988), pp. 277–312; at 277.

2 Harley, "Maps, Knowledge, and Power," 278.

3 Louis O. Quam, "The Use of Maps in Propaganda," *Journal of Geography* 42(1943): 21–32; Louis B. Thomas, "Maps as instruments of propaganda," *Surveying and Mapping*, 9(1949): 75–81.

4 Judith (née Zink) Tyner, *Persuasive Cartography: An Examination of the Map as a Subjective Tool of Communication*, PhD Dissertation, University of California, Los Angeles (1974). See also Judith Tyner, "Persuasive Cartography," *Journal of Geography* 81(1982): 140–144; Judith Tyner, "Images of the Southwest in Nineteenth-Century American Atlases," in *The Mapping of the American Southwest*, Dennis Reinhartz and Charles C. Colley (eds) (College Station: Texas A&M University Press, 1987), 57–77.

5 Norman J.W. Thrower, *Maps and Civilization: Cartography in Culture and Society* (Chicago: University of Chicago Press, 1996), 213.

Discourses, as discussed earlier, refer to *disciplines of knowledge*. Derek Gregory explains that "Discourse refers to all the ways in which we communicate with one another, to that vast network of signs, symbols, and practices through which we make our world(s) meaningful to ourselves and to others."[6] Discourses, therefore, may be understood as frameworks—or, metaphorically, maps. Discourses are ways of ordering and understanding the world. Significantly, this concept of discourse is founded on the ontological position that "reality" itself is produced. As such, discourses do not simply describe the objects of which they refer—they did not have as their correlate an individual or a particular object that is designated by this or that word because there is no "true" referent. There is, for example, no essential *Cambodia* or *Democratic Kampuchea*. This is not to deny the materiality of a "place" called Cambodia; certainly there are bodies of water that have been named "Tonle Sap" and "Phnom Penh." But the *designation* and *meaning* of these "places" is not pre-existing of human thought and practice. When academics or politicians or military strategists refer to Cambodia (or, between 1975–1979 to Democratic Kampuchea) they are not referring to an ontologically pre-existing physical site but instead to a complex political-linguistic construct.

This has important implications, in that maps—as graphic representations—facilitate the production of geographies. Harley, for example, argued that cartography, as a practice, and maps, as social products, are tied into power/knowledge relationships.[7] Maps are not simply objects that represent truth; rather, they are intricately tied into the production of truth. French theorist Michel Foucault asserted that, in politics, there is a battle for "truth." For Foucault, "truth" does not refer to the ensemble of truth that is waiting to be discovered and accepted, but instead to the ensemble of rules, according to which the "true" and the "false" are separated. It is not, therefore, a battle "on behalf" of truth, but rather a battle about what is accepted and circulated—what is "willed" as truth and how this understanding coincides with political, social, and economic relations.[8]

The "will to truth" constitutes part of what Foucault termed a "political economy" of truth. By extension, we may apply these insights to our discussion of maps. Foucault specified some key conditions of how "truth" is willed. First, truth is embedded in particular discourses and the institutions that produce them. Consequently, we may initially surmise that maps as discourse contain within them particular notions of truth. Indeed, maps have long been associated with other dominant discourses, such as that of "science." As such, maps are often described in terms of "accuracy," "preciseness" and "certainty." And what of the institutions that produce maps? The cartographers themselves? Although cartographers well understand that all maps distort—it is not possible to map three-dimensional

---

6 Derek Gregory, *Geographical Imaginations* (Cambridge, MA: Blackwell, 1994), 11.

7 J.B. Harley, "Deconstructing the Map," *Cartographica*, 26(1989): 1–20. See also J.B. Harley, "Cartography, Ethics, and Social Theory," *Cartographica*, 27(1990): 1–23; Denis Wood, *The Power of Maps* (New York: The Guilford Press, 1992).

8 Michel Foucault, "Truth and Power," in *Power/Knowledge: Selected Interviews and other Writings, 1972–1977* (ed.) C. Gordon, translated by C. Gordon, L. Marshall, J. Mepham, and K. Soper (New York: Pantheon Books, 1980), 109–133; at 132.

space onto a two-dimensional place without distortion—there is also a tendency to infer an "objective" intentionality among cartographic practices. In other words, as Tyner explains, map-readers "assume maps are true representations of reality." She concludes, in fact, that "maps are effective in persuasion [because] the general public *accepts* printed maps as "truth" and *expects* them to be "accurate" (emphasis added).[9]

Second, "truth," according to Foucault, is contingent on political, social, and economic demands. Truth is constructed to satisfy particular demands, such as to conform with other policies and practices. Maps, therefore, become crucial in the forwarding of particular political, economic, and social agendas. Maps are used to *persuade* people of the *legitimacy* of certain policies and practices; maps become tools of *justification*. And lastly, for Foucault, "truth" is mobile; it is produced, circulated, and consumed. However, the production, circulation, and even consumption is most often under the control—dominant, if not exclusive—of a few select political and economic apparatuses.[10] As Del Casino and Hanna argue, maps are "not simply representations of particular contexts, places, and times. They are mobile subjects, infused with meaning through contested, complex, intertextual, and interrelated sets of socio-spatial practices."[11] The work of Del Casino and Hanna is especially salient because it emphasizes that "maps stretch beyond their physical boundaries; [maps] are not limited by the paper on which they are printed or the wall upon which they might be scrawled." Instead, maps continue to provide new (re)presentations, new moments of production and consumption, authoring and reading, objectification and subjectification.[12]

After the Communist Party of Kampuchea emerged victorious, *a new map appeared*, one that symbolically spoke to the new country. It is my contention that this map—along with numerous other maps that were proposed by the Khmer Rouge—provides important insights into the place-making strategies that undergirded the mass violence unleashed on the Cambodian people. And in this way, I paraphrase Harley: It is simply not possible to talk about the Khmer Rouge's geographical imagination without encompassing a map.

---

9 Judith Tyner, "Persuasive Cartography," in *History of Cartography* (forthcoming).

10 Foucault, "Truth and Power," 31–32.

11 Vincent J. Del Casino, Jr. and Stephen P. Hanna, "Beyond the 'Binaries': A Methodological Intervention for Interrogating Maps as Representational Practices," *ACME: An International E-Journal for Critical Geographies* 4(2006): 34–56; at 36. See also Matthew Sparke, "A Map that Roared and an Original Atlas: Canada, Cartography, and the Narration of Nation, *Annals of the Association of American Geographers* 88(1998), 463–495; Jeremy Crampton, "Maps as Social Constructions: Power, Communication, and Visualization," *Progress in Human Geography* 25(2001): 235–252.

12 Del Casino and Hanna, "Beyond the 'Binaries,'" 36. See also Tyner, "Images of the Southwest," 58.

A Map of Democratic Kampuchea

> *... the geography of a place is not entirely a result of what actually may be seen there but also of our idea of the place.*
>
> Judith Tyner[13]

It is common for scholars of Cambodia's genocide to provide a map of Democratic Kampuchea's administrative boundaries. It is less common for these scholars to contemplate the significance of the Khmer Rouge's map.

The Khmer Rouge's map portrays the administrative divisions of Democratic Kampuchea. Both David Chandler and Michael Vickery have described, at length, the dimensions of this map, and I borrow liberally from their discussions.[14] Democratic Kampuchea was divided into seven geographic zones, identified by cardinal compass directions: North, Northeast, East, Southwest, West, Northwest, and Center. The Khmer Rouge also identified the Kratie Special Region No. 505 and, prior to mid-1977, the Siem Reap Special Region No. 106. The main seven zones were apparently derived from military designations established by the Khmer Rouge during their war (1970–1975) against the Lon Nol government.

The seven zones did not conform to pre-existing divisions. The Northeast, East, and Southwest zones, all of which bordered Vietnam, included the former eastern portion of Stung Treng province, and the provinces of Ratanakiri, Mondulkiri, Prey Veng, Svay Rieng, eastern Kompong Cham, Kandal, southern Kompong Speu, and Kampot. The Western Zone was composed of Koh Kong, the remainder of Kompong Speu, and Kompong Chhnang. The Northwestern and Northern zones, bordering Thailand, included the provinces of Pursat, Battambang, Oddar Meanchey, Siem Reap, Preah Vihear, and western Stung Treng. The Central zone, lastly, was composed of Kompong Thom, western Kratie, and western Kompong Cham.[15]

The main administrative zones were subdivided into twenty-nine administrative sectors, or *dombon*; these were identified by number. According to Chandler, dombon numbers do not seem to have composed a single numbering system, but rather the traces of several earlier ones, reflecting zonal differences in the wartime period. The actual boundaries of the dombon are unknown, and seem to correspond to those of districts (*srok*) in pre-revolutionary times. Within boundaries, furthermore, villages retained pre-revolutionary names. On the landscape, however, some villages were abandoned, others founded, and many more combined into cooperatives (*sahakor*).[16]

---

13 Tyner, "Images of the Southwest," 59–60.

14 Michael Vickery, "Democratic Kampuchea—Themes and Variations," in *Revolution and its Aftermath in Kampuchea: Eight Essays*, David P. Chandler and Ben Kiernan (eds) (New Haven, CT: Yale University Southeast Asia Studies, 1983), pp. 99–136; David P. Chandler, *The Tragedy of Cambodian History: Politics, War, and Revolution since 1945* (New Haven, CT: Yale University Press, 1991).

15 David P. Chandler, *The Tragedy of Cambodian History: Politics, War, and Revolution since 1945* (New Haven, CT: Yale University Press, 1991), 267.

16 Chandler, *Tragedy of Cambodian History*, 267–269.

Living conditions during Democratic Kampuchea varied greatly over space and scholars have sought to spatially relate specific Khmer Rouge practices with estimates of mortality.[17] In these efforts, the political geography—the various zones, sectors, and districts—imagined by the Khmer Rouge are highly significant. However, the map reveals something *deeper*, more conceptual. Certainly one sees evidence of the militarized society promoted by the Khmer Rouge. The fact that political divisions were derived from military necessity is paramount. However, the map also reveals how the Khmer Rouge sought to erase previous regional identities. Their entire political geographic organization of Democratic Kampuchea was based on an abstract system composed of cardinal direction points and numbers. Geographers, for example, frequently talk of toponyms, or place names. Places are often named after physical or natural features. Place names may indicate what people in a particular location do or believe, or something about that location's history.

But the administrative map of Democratic Kampuchea reflects an attempt to eliminate, or erase, such identifying features. Although such an attempt can be read many ways, I forward one possibility. The Khmer Rouge's cartography signifies "egalitarianism" in that all regions are identical; there is nothing to distinguish one from the other. Consequently, the Khmer Rouge's map reveals significant insight into the *place-making* of Democratic Kampuchea.

In Chapter 4 I argue that the mass violence unleashed in Cambodia by the Communist Party of Kampuchea (CPK) corresponds to a strategy of war communism that sanctioned extreme sacrifice. The Khmer Rouge's total revolution required a total elimination of earlier modes of production. Part-and-parcel of this process was the attempt to provide a degree of uniformity on the country that is unparalleled in modern history. As Chandler concludes, as the CPK "imposed its totalizing policies, it also imposed a consistency of experience on millions of Khmer."[18] In this chapter I consider in more depth how the Khmer Rouge sought to *construct* Democratic Kampuchea as a place.

The Meaning of Place

Tim Cresswell writes, "Place is a word that seems to speak for itself." And it says many things, to many people. Place, for example, may suggest ownership or a connection between a particular person and a particular location (i.e., a house or apartment). Place may also suggest privacy and belonging (i.e., this is *my* place, not yours), or a position in a social hierarchy (i.e., putting someone in their "place"). Place can also suggest emotional, subjective, or even "objective" qualities (i.e., this is a *nice* place, this is an *evil* place, or this is a *warm* place with an average rainfall of

---

17 Craig Etcheson, *After the Killing Fields: Lessons from the Cambodian Genocide* (Lubbock: Texas Tech University Press, 2005); see also Ben Kiernan, *The Pol Pot Regime: Race, Power, and Genocide in Cambodia under the Khmer Rouge, 1975–1979* (New Haven, CT: Yale University Press, 1996).

18 Chandler, *Tragedy of Cambodian History*, 272.

38 centimeters).[19] Given all of these usages of place, how are we to make sense of the word? And what will these have to "say" about mass violence in Cambodia?

Just as the concept of "space" been subject to critical interrogation over the past three decades in Geography and other social sciences, so too has the concept of "place."[20] And similar to space, place is also "both simple and complicated."[21] In fact, as Cresswell explains, place is not a specialized piece of academic terminology and, as such, since we already think we know what it means, it is hard to get beyond that common-sense level in order to understand it in a more developed way.[22] Edward Relph concurs, noting that "We live, act and orient ourselves in a world that is richly and profoundly differentiated into places, yet at the same time we seem to have a meager understanding of the constitution of places and the ways in which we experience them."[23]

Place, not unlike space, was initially conceived by Geographers in absolute terms. Place, for example, was understood as a portion of geographic space; places were unique and were approached as self-contained units.[24] With the rise of humanistic geography, however, there emerged a greater emphasis with a "sense of place." Geographers began to question how emotion, attachment, and subjectivity informed, and were informed by, place. The work of Nicholas Entrikin, for example, sought to mediate between objective and subjective usages of place. This is what Entrikin meant by the "betweenness" of place. Place is both the external context of our actions (hence, objective conditions) but also a center of meaning (people's subjectivity). Any conception of place, therefore, must negotiate these two different ontological positions.

Places were increasingly conceived in a relational sense; places were both contingent and historical, and were constituted of social, political, and economic

---

19 Tim Cresswell, *Place: A Short Introduction* (Malden, MA: Blackwell, 2004), 1.

20 See, for example, Yi-Fu Tuan, *Topophilia: A Study of Environmental Perception, Attitudes and Values* (Englewood Cliffs, NJ: Prentice Hall, 1974); Yi-Fu Tuan, *Space and Place: The Perspective of Experience* (Minneapolis: University of Minnesota Press, 1977); Edward Relph, *Place and Placelessness* (London: Pion, 1976); Anne Buttimer and David Seamon (editors), *The Human Experience of Space and Place* (London: Croom Helm, 1980); John Agnew, *Place and Politics* (Boston: Allen and Unwin, 1987); Nicholas Entrikin, *The Betweenness of Place: Towards a Geography of Modernity* (London: Macmillan, 1991); Robert Shields, *Places on the Margin* (London: Routledge, 1991); Kay Anderson and Faye Gale (editors), *Inventing Places: Studies in Cultural Geography* (London: Belhaven Press, 1992); James Duncan and David Ley (eds), *Place/Culture/Representation* (New York: Routledge, 1993); Marc Augé, *Non-Places: Introduction to an Anthropology of Supermodernity* (London: Verso, 1995); Tim Cresswell, *In Place/Out of Place: Geography, Ideology and Transgression* (Minneapolis: University of Minnesota Press, 1996); Edward Casey, *The Fate of Place: A Philosophical History* (Berkeley: University of California Press, 1998); and Heidi Nast and Steve Pile (eds), *Places Through the Body* (London: Routledge, 1998).

21 Cresswell, *Place*, 1.

22 Cresswell, *Place*, 1.

23 Relph, *Place and Placelessness*, 6.

24 Phil Hubbard, Rob Kitchin, Brendan Bartley, and Duncan Fuller, *Thinking Geographically: Space, Theory and Contemporary Human Geography* (London: Continuum, 2002), 16.

relations. The writings of Ed Relph were (and are) especially provocative. Relph argues that "In our everyday lives places are not experienced as independent, clearly defined entities that can be described simply in terms of their location or appearance. Rather they are sensed in a chiaroscuro of setting, landscape, ritual, routine, other people, personal experiences, care and concern for home, and in the context of other places."[25] From this, it was argued that places do not simply exist, but (rather like space) are produced.

To get a better grip on these re-workings of place, consider the following account provided by Tim Cresswell.

> Cast your mind back to the first time you moved into a particular space—a room in a college accommodation is a good example. You are confronted with a particular area of floor space and a certain volume of air. In that room there may be a few rudimentary pieces of furniture such as a bed, a desk, a set of drawers and a cupboard. These are common to all the rooms in the complex. They are not unique and mean nothing to you beyond the provision of certain necessities of student life. Even these bare essentials have a history. A close inspection may reveal that a former owner has inscribed her name on the desk in an idle moment between classes. There on the carpet you notice a stain where someone has split coffee. Some of the paint on the wall is missing. Perhaps someone has used putty to put up a poster. These are the hauntings of past inhabitation. This anonymous space has a history—it meant something to other people. Now what do you do? A common strategy is to make the space say something about you. You add your own possessions, rearrange the furniture within the limits of the space, put your own posters on the wall, arrange a few books purposefully on the desk. Thus space is turned into place. Your place.[26]

I quote Cresswell at length because his description is rich in imagery but, more saliently, he speaks to several fundamental concepts that Geographers have directed toward the study of place. In fact, it is possible to juxtapose the work of another Geographer, John Agnew, with that of Cresswell. Agnew, in his highly influential book *Place and Politics*, outlines three fundamental aspects of place as a "meaningful location": location, locale, and sense of place. Location, in this sense, refers to the observation that places (in general) always have a fixed objective co-ordinate; places can be located on the Earth's surface, and all places have a unique site.[27] Locale refers to the material setting for social relations. This is based on the observation that all places (again, generally speaking) have a material, or concrete form.[28] This idea parallels that of Henri Lefebvre's concept of "spatial practice," introduced in the previous chapter. Lastly, "sense of place" is used to capture the subjective and emotional attachment people have to place.

If we return to Cresswell's example of the college dorm room, we find all three elements at work. Certainly the room has a specific location—a single and unique

---

25 Relph, *Place and Placelessness*, 29.

26 Cresswell, *Place*, 2.

27 Cresswell notes that some "places" are not always stationary. A ship, for example, may constitute a particular "place" for some people, even though its location is constantly changing. See Cresswell, *Place*, 7.

28 Even imaginary places, Cresswell explains, have an imaginary materiality; fantasy novels, for example, still contain castles and villages, fields and forests.

room. And this room—this space—has certain objective characteristics: its size, shape, and so on. Cresswell then notes that while these dimensions, and those objects initially found in the room (i.e., the desk, the carpet) are common and seemingly devoid of meaning, he also suggests that these objects (indeed, the space of the room) have histories. The spilt coffee, the inscription on the desk: these are, in Cresswell's memorable phrase, the "hauntings of past inhabitation." Thinking back to our earlier discussion of space, and particularly as conceived by Lefebvre, we may surmise that places—such as Cambodia—also contain "hauntings of past inhabitation." For Lefebvre, these hauntings most reflect previous modes of economic production; for Cresswell, however, these entail much more. These histories provide a sense of place, an attachment to a locale. And it implies a sense of possession: turning space into *my* place.[29]

The concepts developed through Cresswell and Agnew's work on place has significance for our understanding of the Khmer Rouge's geographical imaginations. Thus far I have argued that the Khmer Rouge attempted to erase all previous spaces. In effect, they attempted to "exorcize" those "hauntings of past inhabitation." Consider the account provided by Haing Ngor as he describes "life" under the Khmer Rouge: " ... the old way of life was gone and everything about it half forgotten, as if it had never really existed in the first place. Buddhist monks, making their tranquil morning rounds, didn't exist anymore. Three-generation families, where the grandparents look after the little children, didn't exist anymore. Shopping for food in the markets and staying to gossip. Inviting friends over to eat and drink and talk in the evening. It was all gone, and without that pattern we had nothing to hold on to."[30]

The Khmer Rouge sought to transform Cambodia by severing all past attachments to place and, in the process, transform the country into *their place*.

Inside/Outside Democratic Kampuchea

On September 27, 1977 Pol Pot declared: "The enemy is everywhere: in the parliament, the courts, the prisons, the police and the armed forces. Enemy networks honeycomb the country. The make up of social classes in the towns is very complex and varied."[31] According to Locard, these two sentences are extremely important, in that they highlight that "all men, as well as women, children, and the sick from the towns were possibly outcasts in the revolutionary scheme of things." For Locard, this "affirmation contains the germ of logic for the exterminations."[32]

An awareness of group boundaries serves to marginalize and exclude other people. According to Iris Young, a social group is a collective of persons differentiated from at least one other group by cultural forms, practices, or ways of life. More precisely, groups are expressions of social relations; groups only exist in relation

---

29 Cresswell, *Place*, 2.

30 Haing Ngor (with R. Warner), *Survival in the Killing Fields* (New York: Carroll and Graf Publishers, 1987), 247.

31 Quoted in Locard, *Pol Pot's Little Red Book*, 184.

32 Locard, *Pol Pot's Little Red Book*, 184.

to other groups.[33] As Young elaborates, many groups may find themselves defined and marginalized. Indeed, a "whole category of people may be expelled from useful participation in social life and thus potentially subjected to severe material deprivation and even extermination."[34]

Susan Opotow suggests that norms, moral rules, and concerns about rights and fairness govern our conduct toward other people.[35] However, not every person or group is necessarily included within the scope of justice. Rather, she explains that "Inclusion in the scope of justice means applying considerations of fairness, allocating resources, and making sacrifices to foster another's well-being." Conversely, moral exclusion "rationalizes and excuses harm inflicted on those outside the scope of justice. Excluding others from the scope of justice means viewing them as unworthy of fairness, resources, or sacrifice, and seeing them as expendable, undeserving, exploitable, or irrelevant."[36]

The process of social categorization does not proceed, however, based on natural divisions of humanity. Rather than presuming that such categorizations (e.g., racial or ethnic) are simply present, social categories do not simply *include* groups, but rather *produce* groups. In other words, there is no ontological basis to social categories. Furthermore, this production of social groups contributes to a broader form of moral exclusion: the denial of self as manifest in practices of dehumanization. Considerable work on genocides has identified the dehumanizing processes that accompany mass violence. James Waller explains that dehumanizing practices facilitates other practices, such as exclusion, oppression, discrimination and violence.[37] He notes that during the Holocaust, for example, "the Nazis redefined Jews as 'bacilli,' 'parasites,' 'vermin,' 'demons,' 'syphilis,' 'cancer,' 'excrement,' 'filth,' 'tuberculosis,' and 'plague.' In the camps, male inmates were never to be called 'men' but *Haftlinge* (prisoners), and when they ate the verb used to describe it was *fressen*, the word for animals eating. Statisticians and public health authorities frequently would list corpses not as *corpses* but as *Figuren* (figures or pieces), mere things, or even rags. Similarly, in a memo of June 5, 1942, labeled 'Secret Reich Business,' victims in gas vans at Chelmno are variously referred to as 'the load,' 'number of pieces,' and the 'merchandise.'"[38]

There is an immediate *geography* to the process of social categorization. For Waller, linguistic dehumanization is complemented by physical machinations that make the victims seem less human. These degrading, often ritualistic, processes, remake the individual self in the institutional image of something less than a full person.[39] One immediately thinks of the concentration and extermination camps of

---

33 Iris M. Young, *Justice and the Politics of Difference* (Princeton, NJ: Princeton University Press, 1990), 43.

34 Young, *Justice*, 53.

35 Susan Opotow, "Reconciliation in Times of Impunity: Challenges for Social Justice," *Social Justice Research* 14(2001): 149–170.

36 Opotow, "Reconciliation," 156.

37 James Waller, *Becoming Evil: How Ordinary People Commit Genocide and Mass Killing* (Oxford: Oxford University Press, 2002).

38 Waller, *Becoming Evil*, 246.

39 Waller, *Becoming Evil*, 247.

Nazi Germany. As Waller concludes, "Not only do social categorizations systemize our social world; they also create and define our *place* in it" (emphasis added) [40]

According to Cresswell, the construction of places forms the basis for the possibility of transgression. He elaborates that "When something or someone has been judged to be 'out-of-place' they have committed a transgression. Transgression simply means 'crossing a line.'" However, Cresswell explains, "Unlike the sociological definition of 'deviance' transgression is an inherently spatial idea. The line that is crossed is often a geographical line and socio-cultural one. It may or may not be the case that the transgression was intended by the perpetrator. What matters is that the action is seen as transgression by someone who is disturbed by it." [41] Or by someone who has the authority to define what constitutes "out-of-place."

The Khmer Rouge constructed many different populations that were subject to possible exclusion or eradication. Some groups were defined by perceived ethnic divisions: the Viet-Khmer, Sino-Khmer, and Muslim Cham. Other marginalized groups were based on previous occupation or government positions (e.g., soldiers and officials or the Lon Nol government, merchants, capitalists, wealthy farmers and landowners, intellectuals, monks, students, dancers, poets, and musicians).

Most prevalent was the Khmer Rouge's division between "new" and "old" people. This distinction was based not on age, but rather on one's relationship to the revolution. New people—also called "17 April" people or "war slaves"—consisted of both urbanites and rural refugees who had fled to the cities to escape the war. They were differentiated from the "old" people or "base" people who had lived in Khmer Rouge-controlled zones during the war. These were people identified as belonging to the "basic" classes of the poor and lower-middle peasantry. Consider the following slogans used in reference to the "new" people: "The '17 April' are the vanquished of the war"; "The '17 April people' are parasitic plants"; and "The new people bring nothing but stomachs full of shit and bladders bursting with urine." [42]

Both "new" and "old" people found their lives radically transformed in Democratic Kampuchea. [43] According to Locard, "For Khmer Rouge revolutionaries, numerous social groups were liable to be considered superfluous: town dwellers, civil servants, monks, foreigners, and the sick." [44] It was the new people, though, who were especially singled out for even more dehumanizing practices. New people, for example received less food, were treated more harshly, had fewer rights, and were killed more readily than "old" people.

A person's status in Democratic Kampuchea was further differentiated by other factors, including one's "political attitude" and class background. Political attitude, for example, referred to a three-part classification of membership in cooperatives. Persons with full rights, known as *penh sith*, were entitled to join the Party and the

---

40 Waller, *Becoming Evil*, 239.

41 Cresswell, *Place*, 103.

42 Quoted in Locard, *Pol Pot's Little Red Book*, 185. Alexander L. Hinton, *Why Did They Kill? Cambodia in the Shadow of Genocide* (Berkeley: University of California Press, 2005), 9.

43 Hinton, *Why Did They Kill?* 81.

44 Locard, *Pol Pot's Little Red Book*, 181.

army; within the cooperatives, these individuals could hold any political position and receive full rations. Candidates, or *triem*, could hold some low-ranking political positions and were next in line for food distribution. Lastly, there existed a class designated as depositees, or *bannhau*. These persons were generally "last on the distribution lists, first on execution lists, and had no political rights."[45] Under the Khmer Rouge, it was possible to change one's political attitude—depending on one's actions—but class background and the base/new people designations were essentially fixed.

Within the "place" of Democratic Kampuchea, these social divisions created and reinforced a culture of impunity. Hinton explains that "Since 'new people' were less than fully human, there were fewer moral inhibitions in harming them. A "new person" who did something wrong could be "discarded"—a euphemism for execution—without many qualms."[46] For those Khmer Rouge cadre who were responsible for the day-to-day murders, their victims were reduced to a state-of-being less than that of animals; they were considered to be less than human.

Place-Planning of the Khmer Rouge

I ask the Zone to make maps ... [47]

In 1976 Edward Relph published his seminal work *Place and Placelessness*. Relph argued that a practical knowing of places is essential to human existence, and that the basic meaning of place does not come from locations, nor from the trivial functions that places serve, nor from the community that occupies it. Rather, the essence of place lies in the largely unselfconscious intentionality that defines places as profound centers of human existence.[48] It is this relationship between Relph's notion of place and human existence that intrigues me.

Thus far, I have argued that the Khmer Rouge sought to erase the previous spaces of Cambodia, thereby providing a blank slate upon which to construct a new place, that of Democratic Kampuchea. Derived from Marxist understanding of history, this process of *destruction* and *construction* was designed to promote a change in humanity through economic transformation.

Relph explains that "It is not just the identity of a place that is important, but also the identity that a person or group has with that place, in particular whether they are

---

45 May Ebihara, "Revolution and Reformulation in Kampuchean Village Culture," in *The Cambodian Agony*, 2nd edition, David A. Ablin and Marlowe Hood (eds) (New York: M.E. Sharpe, 1990), 16–61; at 25.

46 Hinton, *Why Did They Kill?*, 86.

47 "Excerpted Report on the Leading Views of the Comrade Representing the Party Organization at a Zone Asesmbly," translated by Ben Kiernan, in *Pol Pot Plans the Future: Confidential Leadership Documents from Democratic Kampuchea, 1976–1977*, translated and edited by David P. Chandler, Ben Kiernan, and Chanthou Boua (New Haven, CT: Yale University Southeast Asia Studies, 1988), 9–35; at 34.

48 Relph, *Place and Placelessness*, 43.

experiencing it as an insider or as an outsider."[49] Relph's argument, thus, dovetails readily with the aforementioned discussion of group construction and transgression. Once defined, groups are conceived (often by those in dominant positions) as being "inside" or "outside" an established norm. Relph, however, expands our understanding through a discussion of various levels of intensity of the experience of outsideness and insideness in places. Specifically, Relph develops a seven-fold typology in an attempt to flesh out the various "experiences" of being an insider or outsider.[50] First, there is "existential outsideness," in which all places assume the same meaningless identity and are distinguishable only by their superficial qualities. According to Relph, this is the position that fascinates poets and novelists, who often are intrigued by a "sense of unreality of the world, and of not belonging." A second relationship entails "objective outsideness," in which all places are viewed scientifically and passively. Objective outsideness involves a deep separation of person and place, and has a long tradition in academic geography as well as that of military planning and politics. Such a position reduces places either to the single dimension of location, or to a space of located objects and activities. "Incidental outsideness" constitutes a third relationship between experience and place. Here, places are experienced as little more than backgrounds for activities and thus, the experience of place is even more detached than that of objective outsideness. Relph provides the example of business persons going from city to city merely to attend conferences and meetings. Place is secondary (at best) to the activities at hand; indeed, the identity of the place is little more than that of a background for the conduct of other functions.[51]

A fourth relationship is that of "vicarious insideness, in which places are experienced in a second-hand way." Relph explains that through paintings or poetry, we "enter" into other worlds and other places. Often, feelings toward these "places" are most pronounced when the depiction of a specific place corresponds with our experiences of familiar places. Fifth, "behavioral insideness" involves a deliberate attendance to the appearance of a place. Here, one perceives and conceives of a place as a set of objects, views, and activities. Sixth, "empathetic insideness" occurs when one understands a place to be "rich" in meaning. Such a position demands a willingness to be open to the significance of a place, to "know" and "respect" its symbols. Lastly, according to Relph, "existential insideness" constitutes the most fundamental and "intense" relationship of experience to place. To experience a place as an existential insider, one experiences place without deliberate or self-conscious reflection, yet all the while *knowing* that the place is full of significance. For Relph, this is a insideness that most people experience when "at home": places

---

49 Relph, *Place and Placelessness*, 45.

50 Relph, *Place and Placelessness*, 51–61.

51 Elsewhere I have argued that Southeast Asia, during the Cold War, served as a "surrogate space" in American foreign policy. While the American military-industrial complex was fighting a proxy war against the Soviet Union and China; Vietnam specifically, but Southeast Asia more generally, was *conceived* as a surrogate space; the place of Vietnam, for example, did not matter so much as the fact that America was resisting Communism *some place*. See James A. Tyner, *America's Strategy in Southeast Asia: From the Cold War to the Terror War* (Boulder, CO: Rowman & Littlefield, 2007).

"are lived and dynamic, full with meanings ... that are known and experienced without reflection."[52]

Following Relph, therefore, it becomes possible to distiguish among and between different *experiences* of being an insider or an outsider. Moreover, it is possible to speculate as to how the Khmer Rouge manipulated the *experience of place* through their destruction of Cambodia and subsequent construction of Democratic Kampuchea. Specifically, I draw on two of Relph's relationships: *objectiveness outsideness* and *empathetic insideness*.

As Marxist Leninist revolutionaries, I surmise that the Khmer Rouge leadership approached the revolution from that stand-point of "objective" outsiders. This may strike some as odd, given that the Khmer Rouge must certainly be seen as "insiders" to Cambodia. They were not, one may argue, foreigners who came to Cambodia as colonial occupiers. However, on closer inspection, we may make a counter-argument. Recall from Chapters 2 and 3, most of the Khmer Rouge leadership came from the elite. They had little or no exposure to the day-to-day lives of the peasantry. Accordingly, they did not experience Cambodia as "empathetic insiders"; they were not necessarily privy to the deeply felt emotional bonds between the people and their homes. Furthermore, many of the Khmer Rouge leadership were schooled in French institutions. On the one hand, this further instilled a separation between the revolutionaries and the people of Cambodia. On the other hand, much of the French-based education—including their Marxist-Leninist learning—was *positivist* in orientation. Marx, as detailed in Chapter 4, proposed a scientific and (supposedly) objective interpretation of revolution. I argue that the Khmer Rouge looked upon Cambodia as detached observers; they were the planners and designers of a revolution and of a place. Such a perspective is readily evident in the administrative map of Democratic Kampuchea. It is also found in the documents and policies produced by the Khmer Rouge.

Consider the "excerpted report" of a speech delivered by a "comrade representing the Party Organization" at an Assembly of cadres of the Western Zone in 1976. Once again, the speaker was most likely Pol Pot. This speech, and the accompanying document is important, according to Kiernan, because of its theme of promoting rapid agricultural development without mechanization and its pronouncement of three tons per hectare as a Zone target for rice production[53] Equally important, however, is Kiernan's observation that the Western Zone was "not a peasants' utopia but the plaything of a small group of intellectuals from Phnom Penh."[54] In other words, through this report we can insight into how the Western Zone, in particular, was conceived as a place to be made through revolutionary activity.

According to the speaker we are told that "This [Western] Zone has a very important duty both in building socialism and in defending the country."[55] The Western Zone was also conceived as a model for other zones. Thus, we learn that "Compared to the other Zones, this Zone is the poorest. It is not graced with natural qualities. [It is] the

---

52 Relph, *Place and Placelessness*, 61.

53 "Excerpted Report," 9.

54 "Excerpted Report," 11.

55 "Excerpted Report," 13.

poorest in terms of resources."[56] And that is why it was perceived as most important: "If we take this experience and make it known throughout the country, other places will have to get stronger, will have to get higher production."[57]

The speaker explained next how the seemingly insurmountable difficulties of the zone would be overcome: "From one aspect it would seem that this Zone is poor, because there are mountains everywhere and the soil does not have much of the national fertility. But from another aspect, this Zone has many flat plains, rivers, lakes, ponds, and an extended coastline. These rivers, lakes, ponds, flat plains, and coastline also have many future forms for us. Comparing the difficult and the good aspects, we can see that the future forms are much better. Therefore this Zone is not poor at all; it has a form that can be built up well."[58] In true dialectical reasoning, the speaker has negated the negatives of the zone's geography. Of course, despite the various references to the region's physical geography, its overall topography, there is a complete lack of empirical justification.

Throughout the report, the speaker is dispassionate but optimistic. A series of questions are raised, followed by concise, and seemingly logically answers. The speaker, for example, notes that "The geography of the Zone includes many mountains and forests, few plains and ricelands." Rhetorically, he then asks: "Can we make a Super Great Leap Forward or not?"[59] Answering in the affirmative, the speak explains: "We must look into this problem and resolve it." Reflecting a positive, but staid attitude, the speaker concludes that "We mustn't just look to break even. We must think first of getting enough, and then of getting a surplus as well."[60] I quote at length:

> To break even means that [workers in] the No. 1 sytem (*robob*) get three cans (of rice per day); and No. 2, two-and-a-half cans; in No. 3, two cans; and No. 4, one-and-a-half cans. But we do not want only this much. We want twice as much as this in order to have capital to build up the Zone and the country. We must have enough to eat, that is, 13 *thang* per person per year. If the Zone has 600,000 people, they must eat 150,000 tons of paddy. But we want more than this in order to locate much additional oil, to get ever more rice mills, threshing machines, water pumps, and means of transportation, both as an auxiliary manual force and to give strength to our forces of production. So we must not get just 150,000 tons of paddy. We must get 300,000, 400,000, 500,000 tons just to break even and be able to build socialism and *completely get away from the former system and out of the former period.*[61]

This passage is revealing in that it makes direct reference to the process of replacing one economic system into another. The passage is also informative in that it highlights the militarism of the Khmer Rouge and of how agricultural production was to be "the same as in war." During the war, the Party "raised the principle of attacking

56 "Excerpted Report," 33.
57 "Excerpted Report," 34.
58 "Excepted Report," 14.
59 "Excerpted Report," 20.
60 "Excerpted Report," 20.
61 "Excerpted Report," 20.

wherever we [could] win, wherever the enemy was weak. And the same goes for the economy. We attack wherever the opportunities are greatest."

Significantly, a scientific account of regional variation is provided. According to the document, "in Region A there are two areas: the upper and the lower areas. The lower area is a good area with very great potential. It can grow both rice and vegetables and also has fish. The same goes for Region B. Its rich area is district C. If we produce only two tons of broadcast rice ... this is not an appropriate amount. We must attack so as to get eight tons. For example, on 5,000 hectares of land, the produce is up to 40,000 tones—20,000 to support the 80,000 people, that is, to support all the people in Region B, and there are still 20,000 tons of paddy left. So if we attack on target, just in the one district C, we can feed the whole Region and with half still left over."[62] The document continues that "As for Region D, its rich area is district E. It has 8,000 hectares of good land. If it gets eight tons (per hectare) in two harvests, the produce is up to 64,000 tons. Even though the more than 70,000 people of the whole Region D eat 20,000 tons, there are still more than 40,000 tons of paddy left."[63]

Two final examples will suffice. The Party spokesperson identified that compounding problems, such as sufficient numbers of oxen and buffaloes, required geographic approach. "The problem of insufficient oxen and buffaloes can be resolved if it is organized within the Zone framework. Regions cannot make do on their own, unless the Zone can provide ... 50 or 100 pairs of draught animals. As we know, Region F has many oxen and buffaloes."[64] The report also identifies that "the earth of district E is soft. In district G they have experience in making ploughs with double tips but yoking only one ox. In Region D it is said that eight or ten families have one pair of oxen or buffalo between them. If this is all we have, could we resolve things by using just one ox or buffalo to pull one plough?"[65]

I maintain that this document, as well as many others, reflect a modernistic, pragmatic understanding on the part of the Khmer Rouge. I do not suggest that the Khmer Rouge were correct in their regional economic understanding, but merely emphasize that the Khmer Rouge approached the entire country as "objective outsiders." The Khmer Rouge acted much as "scientists" or "regional planners" might approach their tasks, delimiting the problem, establishing parameters, and identifying solutions. That this was accomplished without empirical justifications should not discount their overall approach. Thus, according to the document, "If we have a great deal of capital, we give some to the Zone and keep some for ourselves for use in our Region. The Zone takes this capital to use to help other places such as district F in Region A. Proceeding in this way, we would get strong quickly. We attack in strong places in order to get capital to help pull up other places which are not favored with resources and lack capital."[66] Furthermore, the document continues:

---

62 "Excerpted Report," 20–21.
63 "Excerpted Report," 21.
64 "Excerpted Report," 22.
65 "Excerpted Report," 22.
66 "Excerpted Report," 28.

In future we will plant rice as well as various other crops. For example, rice is grown in the lower part of Region B. In the higher part we grow green beans and peanuts. So we have both modern agriculture and a variety of products for sale abroad. We have more capital to further expand our Regions and Zone. The same goes for Region G. It looks poor, but in fact it is a rich person's place. If we could resolve the rice problem, keeping enough for consumption and dividing into two parts the quantity left over, one portion is to be given to the Zone, the other is too keep to build up our Region. Besides this we have sea fish and prawns which we can take to sell abroad and gain even more capital to build roads, buy fishing boats and various other machinery.[67]

Of further interest is the specific techniques identified by the Khmer Rouge to carry out these programs. As explained by the speaker,

I ask the Zone to make maps and give the figures for the area of land in each Region and the figures for each sector of the economy: statistics sector by sector, period by period. As for the Regions, I ask them all to have their own maps and statistical tables. If we do this, the Zone Committee can grasp things and so can other people ... Higher levels must also do this. The Center office(s) must also have statistical tables. This figures are not just to have figures on paper. They are to further our leadership. Now I ask the Zone, Region and district ... to have maps ... Each co-operative must have its own map."[68]

A similar position is indicated in the Party's "Four-Year Plan," introduced during the summer of 1976.[69] This documental begins with a description of Democratic Kampuchea as a place: "Our society is basically a collective society ... There are no longer the oppressive characteristics of the old society."[70] From the outset, therefore, it is clear that the Khmer Rouge envisioned Democratic Kampuchea as constituting a new entity; all remnants of the former society have been (in their minds) eliminated.

The Four-Year Plan continues with a geographic comparison of the Cambodian revolution with other revolutions. According to the document, " ... when China was liberated in 1949, the Chinese prepared to end the people's democratic revolution before they prepared to carry out the reform leading to socialism ... Take the example of Korea, liberated in 1945. Not until 1958 did they establish co-operatives throughout the rural areas ... North Vietnam did the same. Now a similar situation aplies in South Vietnam. They need a long period of time to make the transition."[71] Not so in Democratic Kampuchea, or so the leaders argued. Democratic Kampuchea was different. According to the Four-Year Plan, "we have a different character" from China, North Korea, and Vietnam. They extolled: "Our natural characteristics have

---

67 "Excerpted Report," 28.

68 "Excerpted Report," 34.

69 "The Party's Four-Year Plan to Build Socialism in All Fields, 1977–1980," introduced by David Chandler, translated by Chanthou Boua, in *Pol Pot Plans the Future: Confidential Leadership Documents from Democratic Kampuchea, 1976–1977*, translated and edited by David P. Chandler, Ben Kiernan, and Chanthou Boua (New Haven, CT: Yale University Southeast Asia Studies, 1988), 36–119; hereafter referred to as "Four-Year Plan."

70 "Four-Year Plan," 45.

71 "Four-Year Plan," 46.

given us great advantages compared with China, Vietnam, or Africa. Compared to Korea, we also have positive qualities."[72]

Evincing an environmental determinism, the Khmer Rouge maintained that Democratic Kampuchea was inherently superior when compared to other places. However, proper steps were still required to realize their dream of a Communist utopia. Contained within the document are various sections that when combined indicate the particular "place" that was imagined. For example, the Four-Year Plan provides a list of procedures required to build up people's villages:

- There must be maps and diagrams as clear plans to begin with, and start work, step-by-step each year.
- Build a neat, clean, and proper house for each family.
- With watering places for people and for animals.
- Places for animals to live in, have roofs.
- Hygienic toilets.
- Sheds for fertilizers.
- Carpentry workshops.
- Kitchens and eating houses.
- Schools and meeting places.
- Medical clinics.
- Vegetable gardens, large and small. Villages and homes must be located in the tree glades and among all sorts of crops; villages and homes must not have just the sky above and the earth below them.
- Barbers (and hairdressers).
- Rice barn/warehouses.
- A place for tailoring and darning, etc.[73]

According to the Party cadre, all objects were to have a proper place; these were to be clearly specified through the use of maps and diagrams. The Khmer Rouge also provided a list of basic material necessities that would be required, including "clothing, bed supplies (mosquito nets, blankets, mats, pillows) and other materials for common and individual uses: water pitchers, water bowls, glasses, teapots, cups, plates, spoons, shoes, towels, soap, toothbrushes, toothpaste, combs, medicine (especially inhalants), writing books, reading books, pens, pencils, knives, shovels, axes, spectacles, chalk, ink, hats, raincoats, thread, needles, scissors, lighters and flint, kerosene, lamps, etc."[74] There is no apparent rationale to the list, nor to its organization. It reads like a list compiled from "brain-storming" session. It is not hard to imagine a group of individuals—perhaps Pol Pot himself—sitting around a table, tossing out ideas, with an administrative assistant dutifully recording each statement.

And the Khmer Rouge *appeared* to be complete in its assessment. The Four Year Plan, for example, identified the "people's eating regime" which would include (for

---

72 "Four-Year Plan," 46.
73 "Four-Year Plan," 110–111.
74 "Four-Year Plan," 111.

food): "rice, vegetables, fish, meat, preserved fish, salt, fish sauce, soya sauce, etc." and for dessert, "fruits, sugar, cakes, beans, and various things, etc."[75] Significantly, the Khmer Rouge prepared a ration plan. In 1977 workers would receive one dessert every three days; in 1978 dessert rations would be increased to once every two days; and by 1979, workers would receive dessert rations on a daily basis.[76]

The place of Democratic Kampuchea was to be highly regimented and, presumably, more efficient. Here is how the Four-Year Plan described the distribution of food rations: "Organise, nominate and administer people to take responsibility for cooking tasty and high-quality food and deserts; i.e., there must be a separate group, not people taking turns, who are responsible for cooking and making desserts and consider it a high revolutionary duty."[77] The plan similarly provided a schedule for the people's "working and resting regime:"

- Three rest days per month. One rest day in every ten.
- Between ten and fifteen days, according to remoteness of location, for rest, visiting, and study each year.
- Two months' rest for pregnancy and confinement.
- Those under hospitalisation [are considered] according to the concrete situation.

The document explains, furthermore, that "Resting time at home is nominated and arranged as time for tending small gardens, cleaning up, hygiene, and light study of culture and politics."[78]

As Chandler identifies, the Four-Year Plan reads like a set of notes. No suggestions are made of how the people's material needs—limited as they were conceived— were to be met, nor of which needs were more important than others.[79] Apart from the empirical elements, however, these documents are important in that they provide direct insight into the geographies imagined by the Khmer Rouge. The documents are devoid of any qualities that might be considered as displaying an empathetic insideness. The Khmer Rouge who prepared these documents were not interested in how places were to be experienced emotionally, of how different regions might exhibit different meanings and experiences, factors that might affect the functioning of society. Indeed, in their pursuit of conformity, the Khmer Rouge deliberately sought to eliminate the "individualized" or "personalized" attachment to place. There was to be no emotional involvement or attachment, by either the people or by the Khmer Rouge planners.

---

75 "Four-Year Plan," 111.
76 "Four-Year Plan," 111.
77 "Four-Year Plan," 112.
78 "Four-Year Plan," 112.
79 Chandler, "Four-Year Plan," 41.

A Sense of Placelessness

The Khmer Rouge, I have argued, adopted (or exhibited) a detached relationship to the place of Cambodia and its people. Furthermore, the Khmer Rouge explicitly sought to sever the emotional attachments, the sense of place. This has a direct bearing on Relph's central thesis of "place" and "placelessness." It is Relph's contention that a sense of place may be either authentic and genuine, or inauthentic and artificial. Peet explains that "An authentic sense of place involves being inside and belonging to a place (home, hometown, region) as an individual and member of a community, and knowing this without having to reflect on it. Such an authentic and unselfconscious sense of place remains important for it provides an important source of identity for individuals and communities."[80] Conversely, an "inauthentic" sense of place entails a weakening of the identity of places to the point where they not only look alike, but feel alike and offer the same bland possibilities for experience."[81]

The conformity of Cambodia's population is well-established.[82] The Khmer Rouge attempted to promote a singular sameness of people, one based on an idealized imagination of the peasantry. All people were to dress and groom alike. Men and women were to wear traditional black peasant garb; colorful clothing was banned. Hair-styles were also regulated, with short hair imposed on all. People were to not only dress alike, they were to eat the same meals in communal dining halls. Indeed, we may also reconsider the administrative map of Democratic Kampuchea, with all "localized" identity of regions replaced with cardinal directions.

Such conformity relates, ironically, to Relph's thesis of placelessness. As explained by Cresswell, in the modern world we are surrounded by a general condition of creeping placelessness marked by an inability to have authentic relationships to place because the new placelessness does not allow people to become existential insiders.[83] Peet likewise writes that "There is a widespread sentiment that the localism and variety of places and landscapes characteristics of preindustrial societies are being eradicated. In their stead, we are creating "flatscapes," lacking intentional depth, providing only commonplace and mediocre experiences." The irony lies in the observation that for Relph and those who follow his thesis, such a placeless place results from the expansion and penetration of capitalism into all facets of

---

80 Richard Peet, *Modern Geographical Thought* (Oxford, UK: Blackwell, 1998), 51.

81 Relph, *Place and Placelessness*, 90.

82 Most books on the Khmer Rouge provide lengthy discussions on the regimentation of daily-life under the Khmer Rouge. Survivor accounts also provide in-depth descriptions of specific practices. Especially good discussion are found in May Ebihara, "Revolution and Reformulation in Kampuchean Village Culture," in *The Cambodian Agony*, David A. Ablin and Marlowe Hood (eds) (New York: M.E. Sharpe, 1990), 16–61; May Ebihara, "Memories of the Pol Pot Era in a Cambodian Village," in *Cambodia Emerges from the Past: Eight Essays*, Judy Ledgerwood (ed.) (DeKalb, IL: Northern Illinois University Press, 2002), 91–108; and Kalyanee Mam, "The Endurance of the Cambodian Family under the Khmer Rouge Regime: An Oral History," in *Genocide in Cambodia and Rwanda: New Perspectives*, Susan E. Cook (ed.) (New Haven, CT: Yale Center for International and Area Studies, Genocide Studies Program Monograph Series No. 1, 2004), 127–171.

83 Cresswell, *Place*, 44.

society. Indeed, it was proposed by Relph that inauthenticity is the prevalent mode of existence in industrialized, mass societies.[84]

According to Cresswell, the "processes that lead to this [inauthentic sense of place] are various and include the ubiquity of mass communication and culture as well as big business and central authority."[85] Various authors have identified the existence of department stores, supermarkets, and superhighways as "non-places;" terms such as "Disneyification" and "Wal-Martization" have entered our lexicon to capture these ideas. Peet continues that an "inauthentic sense of place is essentially no sense of place, for it involves no awareness of the deep and symbolic significances of places and no appreciation of their identities." He concludes, the "trend is toward an environment of few significant places, a placeless geography, a flatscape, a meaningless pattern."[86]

In their rejection of capitalism, one might surmise that the Khmer Rouge sought to counter the proposed homogenizing effects of globalization, Westernization, and capitalism. However, the Khmer Rouge elite in fact contributed to a sense of placelessness and inauthenticity. The project of the Khmer Rouge, as Marston suggests, while anti-capitalist, was still thoroughly *modern* in orientation. To be sure, the Khmer Rouge destroyed many institutions that we might describe as being modern: banks, currency, and so forth.[87] However, as Marston explains, modernity in the context of Democratic Kampuchea "is best seen in terms of the past it was casting aside. The modernity being fashioned was not Western and it was not technologically developed: those kinds of modernity were explicitly rejected." Marston elaborates that the "rhetoric of the Khmer Rouge involved radical simplication but was not particularly demagogic or anti-intellectual."[88] Indeed, as evidenced by the surviving documents of Democratic Kampuchea, the Khmer Rouge claimed rationality and logic in their actions. This is not surprising, given that many of the Khmer Rouge were in fact intellectuals. As indicated earlier, many had achieved advanced degrees, and many worked as teachers and journalists.

Despite claims to the contrary, Pol Pot and his cadre sought to impose a radical and utopian geographical imaginary. Such a vision, however, led to the same flatscape forewarned by Relph. The mass conformity proposed by the Khmer Rouge differs little from the mass uniformity produced by capitalism's globalizing tendencies. Just as "mass identities" are produced and disseminated through the media channels and especially by advertising, so too were "mass identities" constructed through the slogans and practices of the Khmer Rouge. Both processes—advanced capitalism and the Khmer Rouge's Communist-based utopian revolution—result in the "most superficial identities of place, offering no scope for empathetic insideness and

---

84 See also Peet, *Modern Geographical Thought*, 51.

85 Cresswell, *Place*, 45.

86 Peet, *Modern Geographical Thought*, 51.

87 It should be remembered, however, that the Khmer Rouge did design, and propose to implement at some later date, a new form of currency.

88 John Marston, "Democratic Kampuchea and the Idea of Modernity," in *Cambodia Emerges from the Past: Eight Essays*, Judy Ledgerwood (ed.) (DeKalb: Northern Illinois University Press, 2002), 38–59; at 46.

eroding existential insideness by destroying the basis for identity with places."[89] The Khmer Rouge, through their annihilation of space, their destruction of cities, their severing of social relations, their violent enforcement of conformity and obedience, obliterated all that was *known*. In the process, the Khmer Rouge constructed a "consensus identity that is remote from direct experience," one that is superficial, for it can be changed and manipulated at the will of the leadership.[90]

An "inauthentic attitude to place," Relph argues, entails a technique "in which places are understood to be manipulable in the public interest and are seen only in terms of their functional and technical properties and potentials."[91] Consequently, just as the Khmer Rouge sought to erase the previous geographies of Cambodia, so too did they seek to create a new place that was designed to serve the revolution. This is crucial and bears repeating. Just as people were expected to serve the revolution—with a rifle in one hand and a hoe in the other—so too were *places* to serve the revolution.

The Khmer Rouge, according to Marston, saw themselves as a movement propelled, almost without human agency, by history. He continues:

> They displayed Weberian rationality in that they pursued conscious goals. They condemned the sentimentality of people who were more attached to family than to the revolutionary social agenda. The highest DK leaders probably also saw themselves as acting rationally in the broader sense of acting logically and in accord with universal reason—following a package of values associated with what is embraced when modernity rejects the traditional.[92]

Consequently, as a thoroughly modern project, the policies and practices of the Khmer Rouge were totalizing. And for any highly centralized authority, such as Pol Pot's inner circle, that exerts "considerable control over economic expansion and physical planning," the capacity of those authorities "for place-making or place destruction is immense."[93]

There remains one final component of placelessness that needs elaboration. Peet, following Martin Heidegger, notes that "Inauthenticiy is expressed through the "dictatorship of the *They*," in which the individual is unwittingly governed by an 'anonymous they.'"[94] In otherwords, inauthentic places are characterized by ambiguity and anonymity, secrecy and surveillance. People are disciplined in their attitudes to place, but they are unsure as to where such discipline is located—a theme I discuss further in Chapter 6.

For the people of Democratic Kampuchea, it was not known until 1977—and even thereafter it was a hazy knowledge—who was in charge of the country. Many peasants, for example, believed that Sihanouk had been returned to the throne—a myth that the Khmer Rouge readily exploited. All that was known was that *Angkar*

---

89 Relph, *Place and Placelessness*, 58.
90 Relph, *Place and Placelessness*, 61.
91 Relph, *Place and Placelessness*, 121.
92 Marston, "Democratic Kampuchea," 47.
93 Relph, *Place and Placelessness*, 115.
94 Peet, *Modern Geographical Thought*, 51.

had become the most central "element" in their lives. Marston explains further that "There was always an ambiguity in its usage" and that "this was not incidental to its usage but very much of the essence of what the word signified in practice."[95]

It was unclear for most people of Democratic Kampuchea as to whether *Angkar* was a person, a group, or an organization. According to Locard, ambiguity was at the very heart of the concept of the *Angkar*. He explains that "Concealing the identity of the leaders was indeed the supreme strategem, but how could the leaders induce allegiance, and even affection, among a population from whom they demanded so many inhuman sacrifices?"[96] What was clear, however, was that *Angkar* represented an authority, an authority other than the person who was immediately speaking.[97] All orders and slogans, for example, were issued in the name of the *Angkar Padevoat*, or Revolutionary Organization.[98]

The term *Angkar* itself provided few clues as to what it referred. Linguistically, *Angkar* is derived from the Pali word "anga," meaning a part of the body; it is also related to the Khmer work "angk," which also denotes a structure, or a limb of a body. Furthermore, the term spoke of "mana-filled" objects, or orderly institutions. All of these meanings came into play when discussing the centralized authority of the Khmer Rouge. *Angkar*, on the one hand, frequently signified an organization, a structure that orders society. On the other hand, *Angkar* assumed a religious-status, as it was portrayed as "a quasi-divine entity, comprising both the party leadership and the populace, [one] that should be worshipped by everyone."[99]

Angkar was idealized through the use of kinship idioms. It was, for example, often referred to as the *puk-mae* (the "dad-mom") of Democratic Kampuchea. This is seen, for example, in the Khmer Rouge slogan that extolled *Angkar* as "the mother and father of all young children, as well as all adolescent boys and girls."[100] Consequently, as the "dad-mom" of the people, *Angkar* was conceived as having "true" knowledge and authority. Idealized in songs and poems, *Angkar* was constructed as the benefactor of the people; it cared for, and protected, its children.[101] Thus, according to the Khmer Rouge: "*The* Angkar *is the soul of the revolution*"; "*The* Angkar *is the soul of the motherland*."[102]

As both "dad-mom," "organized authority," and "God," *Angkar* was portrayed as omniscient and omnipotent. It was described as "having the eyes of a pineapple," and purportedly was able to see into people's minds. Indeed, a Khmer Rouge slogan

---

95 Marston, "Democratic Kampuchea," 56.

96 Locard, *Pol Pot's Little Red Book*, 99.

97 Marston, "Democratic Kampuchea," 56. Marston (p. 56–57) explains also that the DK elite avoided reference to individual agency; they avoided, for example, the use of the first-person personal pronoun. Consequently, Marston argues, the claim to be acting outside of individual agency was not only frightening and powerful when encountered by individuals attempting to deal with local officials, it was also a way of evading individual responsibility for action.

98 Locard, *Pol Pot's Little Red Book*, 99.

99 Hinton, "Purity and Contamination," 68.

100 Locard, *Pol Pot's Little Red Book*, 107.

101 Hinton, *Why Did They Kill?* 130.

102 Locard, *Pol Pot's Little Red Book*, 105.

explained that "*The* Angkar *is the people's brain.*" According to Locard, people were not expected to think at all anymore. The *Angkar* proclaimed itself to be not only the voice of the people, but also the brain. *Angkar*, it was claimed, knew all, including people's inner-most thoughts and desires.[103]

This all-knowing function of *Angkar* certainly facilitated their overall control of society. However, it also contributed to the *intended* place-making practices of the Khmer Rouge. Consequently, when we read documents prepared by the Khmer Rouge, outlining their particular geographical imaginations of Democratic Kampuchea— measured in rice yields or dessert allowances—we gain a better understanding of the specific *place* that was being constructed. Relph notes that economics "is not just a matter of production, distribution, and consumption, but a complete way of life."[104] The Khmer Rouge understood this principle very well. Social transformations may be facilitated through economic revolution. Furthermore (as argued in Chapter 4), this "lesson" provided the justification for the brutal practices that annihilated the spaces of previous modes of production.

Herein lies my final argument vis-à-vis place: The Khmer Rouge, operating as objective outsiders, deliberately sought to construct Democratic Kampuchea as a non-place. For the Khmer Rouge, the very absence of a sense of place and an attachment to place was to be sought after as an economic virtue, in that placelessness makes possible the attainment of greater levels of spatial efficiency.[105] This was the place-making process that justified not only the obliteration of everyday landscapes, but the murder of millions of people. For the people of Cambodia, Democratic Kampuchea was an inauthentic place, stripped of meaning and symbolism.

Conclusions

Let us return to the map of Democratic Kampuchea. At first glance, it is straightforward in appearance. Composed of abstract lines, numbers and letters, it appears as a standard, objective graphic representation of political space. A closer look, however, reveals the map to be a statement. Thus, it indicates more than the administrative organization of Democratic Kampuchea; it speaks to the place-making processes inherent to Khmer Rouge practice.

The map is devoid of humanity. There is no localized knowledge, no place-meanings. And this is deliberate. I maintain that this map is a graphic representation of the Khmer Rouge's geographic imagination of Democratic Kampuchea. Built on the former spaces that once held Cambodia, a new, modern, efficient, well-order, structured place was to be constructed. But it was also to be a non-place, a flatscape whereby previously differentiated localities were to be erased and replaced by a

---

103 Locard, *Pol Pot's Little Red Book*, 110. Significantly, this aspect of *Angkar* provides insight into how the Khmer Rouge envisioned the people of Cambodia, and attempted to produce loyal, obedient, and docile subjects. I discuss this in greater detail in Chapter 6.

104 Relph, *Place and Placelessness*, 115.

105 See, for example, Relph, *Place and Placelessness*, 117.

universal conformity. Places were to be *communally* experienced and, as discussed in the next chapter, this experience was to conform with the Khmer Rouge's attempt to eliminate individualism within society.[106]

---

106 The Khmer Rouge, ultimately, were unsuccessful in their attempt to create a non-place. As the accounts of survivors reveal, coupled with the research of scholars such as May Ebihara, the Khmer Rouge were not able to eliminate the Cambodian people's attachment of place. The people of Cambodia were not simply victims of the Khmer Rouge, they were also resisters. Additional work remains. Needed, for example, is a more complete picture of how the people actively sought to maintain their attachment to the landscape of Cambodia and how they articulated a renewed sense of place and, thus of humanity.

Chapter 6

The Political and the Subject

"I am a counterfeit revolutionary, in fact I am an agent of the enemy, the enemy of the people, and the nation of Kampuchea, and the Communist Party of Kampuchea. I am the cheapest reactionary intellectual disguised as a revolutionary."[1] So confessed Hu Nim under severe torture. Hu Nim served as minister of information and propaganda under the Khmer Rouge. He was also a longtime associate of Khieu Samphan. On April 10, 1977 Hu Nim was arrested and imprisoned at Tuol Sleng, a secret institution run by the Khmer Rouge, a place where approximately 14,000 men, women, and children were detained, tortured, and killed. For nearly three months, until his execution on July 6, 1977, Hu Nim was subjected to brutal forms of torture. And during his stay in captivity, Hu Nim produced over 200 pages of handwritten confessions.

Much has been written about Hu Nim and his forced confessions.[2] David Chandler, for example, maintains that Hu Nim's confession foreshadowed a series of arrests and purges that occurred in Democratic Kampuchea throughout 1977 and 1978. His confession, and the subsequent purges, "constituted a classic case of scapegoating by the Party Center."[3] My interest, however, deviates from that of Chandler and others, in that I am struck by Hu Nim's admission of a "false identity." Who was Hu Nim? Or, more broadly, who was *anyone* under the Khmer Rouge?

"Who are we?" This singular and seemingly straight-forward question dominates the writings of Michel Foucault. Whether Foucault was writing about prisons, medicine, or sexuality, threads of this question were woven through his various projects whereby he sought to "to create a history of the different modes by which … human beings are made subjects."[4] But, as Todd May explains, this question (and overall project)—which certainly was not unique to Foucault—is important because it is so basic, some common-place.[5] Indeed, May asks, "Who has not, at least once

---

1 "Planning the Past: The Forced Confessions of Hu Nim," translated by Chanthou Boua and introduced by Ben Kiernan, in *Pol Pot Plans the Future: Confidential Leadership Documents from Democratic Kampuchea, 1976–1977*, David P. Chandler, Ben Kiernan, and Chanthou Boua (eds) (New Haven, CT: Yale University Southeast Asia Studies, 1988), 227–317; at 240.

2 David P. Chandler, *Voices from S-21: Terror and History in Pol Pot's Secret Prison* (Berkeley: University of California Press, 1999), 64.

3 Chandler, *Voices from S-21*, 65.

4 Michel Foucault, "The Subject and Power," in J. Faubion (ed.) *Power: Essential Works of Foucault, 1954–1984*, Vol. 3, translated by R. Hurley et al. (New York: The New Press, 2000), 326–348; at 326.

5 Todd May, *The Philosophy of Foucault* (Montreal: McGill-Queen's University Press, 2006), 2.

or twice, asked this question? And, when asked seriously, rigorously, its answer—or its response—is not obvious."[6]

In *The Archaeology of Knowledge* Foucault quips, "Do not ask who I am and do not ask me to remain the same: leave it to our bureaucrats and our police to see that our papers are in order."[7] Foucault's statement is complex, referring on the one hand to his contention that we, as humans, change; we are always in a process of becoming. On the other hand, Foucault intimates that there are, however, many apparatuses that do in fact attempt to "fix" us into pre-determined slots; that people are to be regulated, with papers in order, thereby establishing some degree of routinized organization. And we are routinely being told "who we are." As May identifies, "Our church tells us: you are a child of God. Our politicians tell us: you are an American (or an Australian, or an Indian, or …). Our televisions tell us: you are a consumer." He concludes that "We are told who we are, and as a result we rarely ask."[8]

The question of "who we are" resonates with an ongoing theme of *The Killing of Cambodia*, namely to question the common-sense understandings of space and place as manifest in an environment of mass violence (i.e., genocide). However, Foucault's question is important for our present purpose because it directs our attention to the dialectics of "self" and "space/place." Indeed, an understanding of "who we are" is inseparable from the question of "where we are." Foucault's concern resonates with that of Frantz Fanon, who asked: "Where am I to be classified? or, if you prefer, tucked away?"[9] Fanon's question is important because it suggests that *who we are* is a function of *where we are*. In other words, the concerns addressed by both Foucault and Fanon pivot on the idea that certain spaces/places are acceptable for certain people—or bodies—while other spaces/places are not.

The "body" has in recent years emerged as a crucial site of inquiry. Linda McDowell, for example, contends that the body has become a major theoretical preoccupation across the social sciences. David Harvey, likewise, has identified that the "extraordinary efflorescence of interest in the 'body' [has served] as a grounding for all sorts of theoretical enquiries over the last two decades."[10]

How, though, are we to approach the "body"? Both "social theorists" and "natural scientists" have grappled with the "body." On the one hand, "science" traditionally has located the *material body* within the realm of physical, biological, and chemical processes. Consequently, all physical, psychological, and social phenomena, including bodily motions, consciousness, and memory are explained on the basis of the natural sciences. On the other hand, social and cultural theorists have explained the body as socially constructed.[11]

---

6 May, *Philosophy of Foucault*, 2.

7 Michel Foucault, *The Archaeology of Knowledge and the Discourse on Language* (New York: Pantheon Books, 1972), 17.

8 May, *Philosophy of Foucault*, 2–3.

9 Frantz Fanon, *Black Skin, White Masks* (New York: Grove Press, 1967), 113.

10 Linda McDowell, *Gender, Identity and Place: Understanding Feminist Geographies* (Minneapolis: University of Minnesota Press, 1999), 36; David Harvey, *Spaces of Hope* (Berkeley: University of California Press, 2000), 97.

11 Olaf Kuhlke, *Representing German Identity in the New Berlin Republic: Body, Nation, and Place* (Lewiston: Edwin Mellen Press, 2005), 57.

For Olaf Kuhlke, the central problem of both "scientific" and "constructionist" approaches to the body is that they tend to be mutually exclusive; both claim to provide the sole explanatory framework for an understanding of the body. Accordingly, we only think of the body in either/or terms: it is *either* a natural, biological, genetic entity, *or* it is purely cultural, socially constructed, and the product of competing interests and ideologies. Arguing against this false dualism, Kuhlke suggests a third possibility, namely an understanding of the body as *both* thoroughly natural and cultural. This requires us to comprehend the body in both its pre-discursive nature and its discursive construction.[12]

As starting point, Kuhlke turns to the writings of Henri Lefebvre. In Chapter 4 I discussed Lefebvre's contributions to the production of space and, in particular, his forwarding of the concepts of *spatial practice, representations of space*, and *representational spaces*. Spatial practices, it will be recalled, include those routines of daily life that both form and are formed by particular modes of economic organization. Here we consider his conceptualization of the body and, in particular, his idea of the spatially *lived* body. Lefebvre writes:

> Considered overall, social practice presupposes the use of the body: the use of the hands, members and sensory organs, and the gestures of work as of activity unrelated to work. This is the realm of the *perceived* (the practical basis of the perception of the outside world, to put it in psychology's terms). As for *representations of the body*, they derive from accumulated scientific knowledge, disseminated with an admixture of ideology: from knowledge of anatomy, of physiology, of sickness and its cure, and of the body's relations with nature and with its surroundings or 'milieu.' Bodily *lived* experience, for its part, may be both highly complex and quite peculiar, because 'culture' intervenes here, with its illusory immediacy, via symbolisms ... [13]

For Kuhlke, this passage is revealing, in that it highlights the body as something that is inherently spatial; the body is our medium of having a world. The body, for example, is our natural, material substance, our practical tool for navigating through this world. It is, Kuhlke argues, "physical space and it *gives us* space and place." As such, given certain bodily pre-givens—the material body that we are born with—we are positioned in the world, and are enabled or disabled to perform certain tasks, create certain meanings, and attach these to place. This, then, constitutes the *pre-discursive realm of the body*.[14]

However, Kuhlke rightly suggests that "the ability or inability of our physical bodies is continuously negotiated in social contexts." Subsequently, our material bodies are constrained or enabled to do certain things, based on socially constructed norms or regulations. Kulke elaborates that "through the social construction of norms and values, certain biological bodies are conditioned, stigmatized, or stereotyped. Because of our enduring, permanent bodily dispositions and existing socially constructed situations—in which certain norms or conduct of standards of behavior apply—evaluations of the relationship between our disposition and the situation take

---

12 Kuhlke, *Representing German Identity*, 58.

13 Henri Lefebvre, *The Production of Space* (Oxford, UK: Blackwell, 1991), 40.

14 Kuhlke, *Representing German Identity*, 59.

place, and subsequently categorizations of dispositions versus situations are used to evaluate the appropriateness or standard of our behavior."[15] As McDowell concludes, "While bodies are undoubtedly material, possessing a range of characteristics such as shape and size and so inevitably take up space, the ways in which bodies are presented to and seen by others vary according to the spaces and place in which they find themselves."[16]

Lefebvre's writing resonate with those of Michel Foucault. Foucault forwarded an ontological position that critiqued biological essentialism and the belief in a transcendental subject. He argued that "the individual is not a pre-given entity which is seized on by the exercise of power. The individual, with his [sic] identity and characteristics, is the product of a relation of power exercised over bodies, multiplicities, movements, desires, forces."[17] Foucault thus does not deny the materiality of the body; but neither does the body's materiality exist outside a disciplinary framework in terms of both knowledge and practices.[18]

Victoria Pitts, herself influenced by Foucault, explains that many contemporary theorists "reject such notions as the body's universality, naturalness, and subordinate relationships to a rational actor as deeply logocentric."[19] She elaborates:

> Post-essentialist theories of the body, expressed in cultural studies, feminism, postmodernism, poststructuralism, and other areas of thought, reject the notion that there is an 'essential,' proper, ideal body. Instead, the body, along with social laws, nature, and the self, is seen as always open to history and culture, and always negotiable and changing. Instead of one truth of the body or of ontology, there are competing truths that are productions of time, space, geography, and culture.[20]

The body, therefore, has a material existence. However, this physicality does not pre-determine "identity." Who "we" are is not pre-discursive. Wolfgang Natter and John Paul Jones contend that "as products of hegemony, the categories we take as materially significant not only lack a 'natural' basis for grounding identity, but are the very grounds by which identity is produced." Bodies, for example, are conditioned, stigmatized, and stereotyped, and thus become *raced, gendered,* and *sexed.* Natter and Jones elaborate that "It is not just that categories are 'social' ... but rather that those aspects of alterity that are seized upon and amplified into a system of social differentiation are always contingently productive of subjects in the interest of hegemonic power."[21] It is for this reason I employ the term "subject" rather than

---

15 Kuhlke, *Representing German Identity*, 61.

16 McDowell, *Gender, Identity and Place*, 34.

17 Foucault, "Questions on Geography," in C. Gordon, *Power/Knowledge: Selected Interviews and Other Writings, 1972–1977* (New York: Pantheon Books, 1980), 63–77; at 73–74.

18 M.A. McLaren, *Feminism, Foucault, and Embodied Subjectivity* (Albany: State University of New York Press, 2002), 15.

19 Victoria Pitts, *In the Flesh: The Cultural Politics of Body Modification* (New York: Palgrave Macmillan, 2003), 29.

20 Pitts, *In the Flesh*, 28.

21 Wolfgang Natter and John Paul Jones III, "Identity, space, and other uncertainties," in *Space & Social Theory: Interpreting Modernity and Postmodernity*, Georges Benko and Ulf Strohmayer (eds) (Oxford, UK: Blackwell, 1997), 141–161; at 146–47.

"identity." As Catherine Belsey explains, the term "subject" first places the emphasis squarely on the language learned, and from which "subjects" internalize the meanings their society expects them to live by. Second, 'subject' reflects its own ambiguities, its double meanings.[22] As Michel Foucault writes, there are (at least) two meanings of the word "subject": "subject to someone else by control and dependence, and tied to his [sic] own identity by a conscious or self-knowledge. Both meanings suggest a form of power that subjugates and makes subject to."[23] Third, "subject" allows for discontinuities and contradictions, whereas "identity" implies sameness; subjects, however, can differ even from themselves.

This notion of the "subject" differs from the more traditional, Cartesian-based "individual" of the Enlightenment. The latter concept, "individualism," dates from the Renaissance and presupposes that humans are free, intelligent entities, and that decision-making processes are not coerced by historical, political, or cultural circumstances. As Jenny Edkins explains, the "Enlightenment subject was a unified individual with a center, an inner core that was there at birth and developed as the individual grew, while remaining essentially the same. This core of the self was the source of the subject's identity."[24] The poststructural subject, conversely, has no fixed, essential, or permanent identity. "Subjectivity is formed and transformed in a continuous process that takes place in relation to the ways we are represented or addressed and alongside the production or reproduction of the social."[25]

The work of Natter and Jones compliments the approaches adopted by Lefebvre and Kuhlke, namely a recognition of both the *materiality* of the body, but also those *discursive inscriptions* that give meaning to the body. Natter and Jones explain that "To focus on the contingent social construction of the category ... might for some seem to imply an inattentiveness to the material effects ... of that construction. Precisely the opposite is true: it is only through the linkage of both—the construction of the category (i.e., race *as* representation) and the material effects that conform to and reproduce the category—that we can deconstruct with full force the deep structures that construct difference as meaningful and deploy it in hegemonic projects."[26]

By employing a post-structural conception of the body as subject, I argue that the Khmer Rouge in fact exhibited a more modern, "Enlightened" conception of bodies in their subjugation of people to brutal practices in order to eliminate individualism.[27] From the perspective of Pol Pot and his colleagues, the *construction* of Democratic Kampuchea was predicated on the *destruction* of Cambodia. Part and parcel of these spatial practices was the creation of a new political subject, one who would inhabit

---

22 Catherine Belsey, *Poststructuralism: A Very Short Introduction* (Oxford: Oxford University Press, 2002), 52.

23 Michel Foucault, "Subject and power," 331.

24 Jenny Edkins, *Poststructuralism and International Relations: Bringing the Political Back In* (Boulder, CO: Lynn Rienner Publishers, 1999), 21.

25 Edkins, *Poststructuralism and International Relations*, 22.

26 Natter and Jones, "Identity, space, and other uncertainties," 147.

27 Academically, this is a crucial point. It is becoming common-place for researchers to argue (rightly so, I might add) that bodily integuments such as "gender" and "sexuality" are socially constructed. However, researchers must also acknowledge and recognize that many individuals, especially outside of academia, disagree with such sentiments.

the new state. In this chapter I consider how space, place, political subjectivity, and sovereignty interact within the context of mass violence.

Disciplined Bodies

Be careful—bodies disappear.

[Saying during the genocide][28]

Thoroughly modern in orientation, bodies—much as space—for the Khmer Rouge were predicated on a Cartesian understanding. The body, as conceived by the Khmer Rouge leadership, was one of an immutable, fixed essence. This point cannot be overemphasized. It was precisely this ontological conception of the body that legitimated—in the minds of the Khmer Rouge—the ensuing brutal practices that were enacted. Indeed, such an ontological understanding of humanity precluded any widespread practice of political re-education. In conforming with the ideology of erasing all vestiges of previous societies, the Khmer Rouge viewed certain populations and certain bodies as "good" or "bad," "pure" or "impure;" such categorization would directly indicate how bodies were to be used—or eliminated.

Biological bodies were posited as natural and largely unchanging. However, the Khmer Rouge also forwarded a position that conformed with a classical Marxist understanding, namely that bodies as a whole *were* shaped by the existent economic organization of society. In other words, those bodies shaped by capitalism were impure and evil; those that were shaped by socialism, conversely, were pure and good. And not surprisingly, the Khmer Rouge viewed children, most of all, as malleable to change.

A traditional saying in Cambodia holds that "Clay is molded while it is soft."[29] According to Henri Locard, this slogan was often used to signify that only young children could be selected by the Party to become the docile servants of the *Angkar*. Another slogan, reminiscent of Mao's "blank page," claimed that "Only a newborn is free from stain."[30] This idea, in fact, was developed by Pol Pot, who said of the young: "Those, among our comrades, who are young, must make a great effort to re-educate themselves. They must never allow themselves to lose sight of this goal. You have to be, and remain, faithful to the revolution. People age quickly. Being young, you are at the most receptive age, and capable to assimilate what the revolution stands for, better than anyone else."[31] Taken together, these phrases provide insight into the embodied ontology of the Khmer Rouge.

---

28 Haing Ngor (with R. Warner), *Survival in the Killing Fields* (New York: Carroll and Graf Publishers, 1987), 252. According to Ngor, this was a saying that sprang up among the "new" people as a warning not to attract attention.

29 Henri Locard, *Pol Pot's Little Red Book: The Sayings of Angkar* (Chiang Mai, Thailand: Silkworm Books, 2004), at 143.

30 Locard, *Pol Pot's Little Red Book*, 143.

31 Locard, *Pol Pot's Little Red Book*, 144. In anticipation of a later argument: If children were so malleable, why did the Khmer Rouge kill so many babies and children? The answer is complex and relates, on the one hand, to a Khmer conception of disproportionate revenge.

As discussed in Chapter 5, in Democratic Kampuchea the Khmer Rouge routinely (and somewhat arbitrarily) constructed social groups: "new people" and "old people," capitalists, landlords, intellectuals, monks. Certain bodies, therefore, were constructed as *different*. Difference, though, within Democratic Kampuchea was the antithesis of *equality*. Consequently, the Khmer Rouge attempted to achieve a complete state of conformity. All *subjects* were to be alike, all were to be *identical*. In short, an all-pervasive sameness—of dress and demeanor, thought and emotion— was imagined by the Khmer Rouge.

As the Khmer Rouge attempted to promote a singular sameness of people, one based on an idealized geographical vision of the peasantry, all people were to dress and groom alike. Men and women were to wear traditional black peasant garb; hair-styles, likewise, were regimented. Beyond these surface appearances, however, the Khmer Rouge attempted to implement a complete and total subjugation of the body and the mind. All aspects of individualism were to be destroyed. People of Democratic Kampuchea were to lead simple rural lives based on equality and self-sacrifice. Everything was to be done for the masses and for the nation.[32] People's actions were no longer based on individual profit or self-fulfillment; rather, a selfless dedication to the collective well-being of Democratic Kampuchea was promoted. Efforts were made to remove all incentives for individual accomplishment to teach each person that any deviation from the general party line—any selfish or individual action—would result in severe punishment and probable death.[33]

What Pol Pot and his cadre sought to achieve was the obliteration of individualism from the collective Cambodian psyche. By destroying every vestige of individualist thought, the Khmer Rouge envisioned a new society that would consist of people completely dedicated to, and knowing only, a collectivist regime. Ideologically, the key was to make a utopian society—more quickly and more completely—than any previous attempt, including both the Russian and Chinese revolutions. The intention was not to remake the old society, or to return to the glorious days of Angkor, but to make society anew. This required an explicit attempt to wipe clean all that existed before the revolution. Instantaneously, every member of Khmer society was to occupy the same economic and social level; contradictions between the wealthy and the poor, the educated and the illiterate, rural and urban, were to be wiped out.[34]

Such conformity required a total and complete system of discipline. Discipline, according to Foucault, is a form of power, "a modality for its exercise, comprising a whole set of instruments, techniques, procedures, levels of applications."[35]

This idea is captured in the saying that to remove weeds, one must pull them out by the root. In other words, to kill one's enemy, you must also kill his/her relatives. On the other hand, the killing of infants and children suggests that not all children were considered blank pages; some were thought to be impure because of their parentage.

32 Nic Dunlop, *The Lost Executioner: A Journey to the Heart of the Killing Fields* (New York: Walker and Company, 2005), 156.

33 Kenneth M. Quinn, "The Pattern and Scope of Violence," in *Cambodia, 1975–1978: Rendezvous with Death*, Karl D. Jackson (ed.) (Princeton: Princeton University Press, 1989), 179–208; at 193.

34 Quinn, "The Pattern and Scope of Violence," 192–193.

35 Foucault, *Discipline and Punish*, 215.

Disciplinary techniques, moreover, whether appropriated by prisons, schools, the military or any other apparatus, share a number of characteristics. First, disciplinary techniques are dual-sided in that they simultaneously enable and repress, organize and atomize.[36] Foucault thus counters the traditional notion that discipline (or power) is a purely negative process, of one individual or institution exerting its dominance over another. Second, disciplinary techniques render each individual a "case." Third, discipline is manifest in rules.[37]

Most salient, however, is the observation that discipline is directed toward bodies. As Foucault writes, "it is always the body that is at issue—the body and its forces, their utility and docility, their distribution and their submission."[38] Foucault continues that "the body is … directly involved in a political field; power relations have an immediate hold upon it; they invest it, mark it, train it, torture it, force it to carry out tasks, to perform ceremonies, to emit signs."[39] Accordingly, discipline is meted on the body and this proceeds, initially, from the distribution of bodies in space.[40]

Disciplines requires enclosure.[41] We have seen, in earlier chapters, how the Khmer Rouge evacuated cities and relocated people into collectives. And within these collectives, people were further segregated by age, sex, and "political" classification. In total, the spatial separation of daily life was augmented with respect to social relations; these spatial separations, therefore, facilitated that severance of traditional familial bonds and redirected loyalties to *Angkar*. As explained by Teeda Butt Mam, who was fifteen years-old when Cambodia ceased to exist, the Khmer Rouge "kept moving us around, from the fields into the woods. They purposely did this to disorient us so they could have complete control. They did it to get rid of the "useless people." Those who were too old or too weak to work. Those who did not produce their quota."[42] Moly Ly, likewise, survived the Khmer Rouge. He remembers that he, along with his family, was sent to work on a collective in Battambang Province. Within his community, people were divided into five classes: small children, bigger children, single women and men, married women and men, and elderly women and men. He explains that each class was forced to work in a different field. As a teenager, he worked in a *kong Chalat*, or mobile work brigade. His duties included building dikes, digging canals, liquidating the forest by removing roots, chopping logs and branches, and setting old brush on fire.[43]

---

36 R. Marsden, "A Political Technology of the Body: How Labour is Organized into a Productive Force, *Critical Perspectives on Accounting* 9(1998): 99–136; at 120.

37 Marsden, "Political Technology," 121.

38 Foucault, *Discipline and Punish*, 25.

39 Michel Foucault, *Discipline and Punish*, 25.

40 Foucault, *Discipline and Punish*, 141.

41 Foucault, *Discipline and Punish*, 141.

42 Teeda Butt Mam, "Worms From Our Skin," in *Children of Cambodia's Killing Fields: Memoirs by Survivors*, compiled by Dith Pran and edited by Kim DePaul (New Haven, CT: Yale University Press, 1997), 11–17; at 14.

43 Moly Ly, "Witnessing the horror," in *Children of Cambodia's Killing Fields: Memoirs by Survivors*, compiled by Dith Pran and edited by Kim DePaul (New Haven, CT: Yale University Press, 1997), 57–65.

The experiences of Moly Ly and Teeda Butt Mam highlight Foucault's corollary thesis that "enclosure" is neither constant nor sufficient. Rather, "enclosure" must be coupled with "partitioning." According to Foucault, "Each individual has his [sic] own place; and each place its individual." He explains that "Disciplinary space tends to be divided into as many sections as there are bodies or elements to be distributed. One must eliminate the effects of imprecise distributions, the uncontrolled disappearance of individuals, their diffuse circulation, the unusable and dangerous coagulation." Combined, "enclosure" and "partitioning" function as tactics of "anti-desertion, anti-vagabondage, anti-concentration." The aim is to "establish presences and absences, to know where and how to locate individuals." These are, in effect, procedures "aimed at knowing, mastering and using."[44]

Within the work brigades and collectives, individual bodies were under constant supervision; their activities were monitored and, at times, recorded. Any movements, any attitudes, any thoughts that were "out of order," that did not conform, were subject to immediate and brutal punishment. Bodies within Democratic Kampuchea remained partitioned, each sequestered into their own (often) enclosed spaces; and through their performance of specific functions, no member was able to see the totality of the system.

Roeun Sam was 14 years-old when the Khmer Rouge came to power.[45] She recalls:

> They put me with others my age and had me work in the field to watch the cows. Every day I watched the cows, and after I fed them at night I went to the place where the children laid on the ground to sleep. We didn't have a roof, wall, or bed. We only ate one meal a day, at lunch. Angka measured each serving, only about a cup and a half, which was mostly broth and maybe two tablespoons of rice.[46]

In her narrative, Sam describes the partitioning of Khmer society, of how different bodies—classified by age and sex—were assigned to different and specific places. She also discusses the fear of going to *other* places, places that were forbidden:

> One day ... I noticed that a few of my cows were missing. I smelled something like a dead animal. My cows were running toward the smell, and I followed them. By the time I got there the cows were licking the dead body's clothes. Some were standing there sniffing. It was a human body that had just been killed. You could see her long black hair and the string around her hands. I looked around and saw people who had been shot and their heads were smashed in. There were at least one hundred people dead. This was the place they took people to kill.[47]

Within the collectives, communal dining halls served a similar purpose. As May Ebihara explains, the "imposition of communal dining halls was not simply a means whereby the state controlled the distribution; it further demonstrated that the work

44 Foucault, *Discipline and Punish*, 143.
45 Roeun Sam, "Living in Darkness," in *Children of Cambodia's Killing Fields: Memoirs by Survivors*, compiled by Dith Pran and edited by Kim DePaul (New Haven, CT: Yale University Press, 1997), 73–81.
46 Sam, "Living in Darkness," 74.
47 Sam, "Living in Darkness," 74–75.

team or cooperative had superseded the family as the basic social unit in Democratic Kampuchea."[48] Kalyanee Mam explains that "collective dining was enforced ... because the regime feared that allowing familes to produce their own food would encourage family interests and distract loyalty from *Angka*. As with other policies implemented by the regime, the purpose of collectivizing food and property was to eliminate individual dependency on the family and [to] force individuals to project this dependency towards the organization."[49]

However, even the enclosure and partitioning of bodies is insufficient in disciplining space. As Foucault identities, "the exercise of discipline presupposes a mechanism that coerces by means of observation; an apparatus in which the techniques that make it possible to see induce effects of power."[50] In Democratic Kampuchea, that apparatus was *Angkar*.

Foucault suggests that "the perfect disciplinary apparatus would make it possible for a single gaze to see everything constantly."[51] He continues that "Disciplinary power ... is exercised through its invisibility; at the same time it imposes on those whom it subjects a principle of compulsory visibility. In discipline, it is the subjects who have to be seen. Their visibility assures the hold of the power that is exercised over them. It is the fact of being constantly seen, of being able always to be seen, that maintains the disciplined individual in his [sic] subjection."[52] In Chapter 5 I indicated that *Angkar*—portrayed as the "dad-mom"—of the revolution, was portrayed as omniscient and omnipotent. It was all-seeing, all-knowing, and all-powerful. The slogan "The *Angkar* has the eyes of a pineapple" thus served as a warning. Locard explains that "In the political sphere, we ... have here a Khmer Rouge version of Orwell's famous aphorism: 'Big Brother is watching you.' But the Khmer Rouge leadership was more wily, since the *Angkar* had no face and there were no posters of Pol Pot on village or town walls."[53] Since no one neither knew who or what *Angkar* was, nor what *Angkar* looked like, its disciplinary function was augmented. It was known, however—and again reminiscent of Orwell's nightmarish world of *Nineteen Eighty-Four*—that *Angkar* enlisted children to serve as its minions. As a young child, for example, Ouk Villa was "taught to call our parents 'comrades' and to spy on them."[54]

---

48 May Ebihara, "Revolution and Reformulation in Kampuchean Village Culutre," in *The Cambodian Agony*, 2nd edition, David A. Ablin and Marlowe Hood (eds) (New York: M.E. Sharpe, 1990), 16–61; at 60.

49 Kalyanee Mam, "The Endurance of the Cambodian Family Under the Khmer Rouge Regime: An Oral History," in *Genocide in Cambodia and Rwanda: New Perspectives*, Susan E. Cook (ed.) (New Haven, CT: Yale Center for International and Area Studies: Genocide Studies Program Monograph Series, No. 1, 2004), 127–171; at 143.

50 Foucault, *Discipline and Punish*, 170–171.

51 Foucault, *Discipline and Punish*, 173.

52 Foucault, *Discipline and Punish*, 187.

53 Locard, *Pol Pot's Little Red Book*, 113.

54 Ouk Villa, "A Bitter Life," in *Children of Cambodia's Killing Fields: Memoirs by Survivors*, compiled by Dith Pran and edited by Kim DePaul (New Haven, CT: Yale University Press, 1997), 115–121; at 116.

Within Democratic Kampuchea, the Khmer Rouge sought to produce a highly disciplined, ranked society. However, even the act of being watched was insufficient. Indeed, as Strub identifies, "the reason mere observation might induce coercive effects of power is that those being observed expect negative consequences to follow the detection of inappropriate behavior. These negative consequences are manifest in punishment, torture, and death."[55] Foucault asserts that public executions have juridico-political functions; that executions are ceremonies by which a momentarily injured sovereignty is reconstituted. He elaborates that:

> The public execution ... deploys before all eyes an invincible force. Its aim is not so much to re-establish a balance to bring into play, as its extreme point, the dissymmetry between the subject who has dared to violate the law and the all-powerful sovereign who displays his [sic] strength.[56]

The torture and execution of bodies should not be viewed as extreme expressions of irrational violence or rage. Rather, torture itself should be conceived as a technique.[57] Foucault maintains that torture forms part of a ritual, one that includes two components. First, it must mark the victim; it is intended to physically or symbolically brand the victim with guilt. Second, the public torture and execution must be spectacular. It must be seen by all as its triumph.[58]

Within Democratic Kampuchea, the ritualized practice of torture was frequently enacted at re-education or self-criticism sessions. According to Locard, self-criticism and mutual criticism constituted basic mental exercises in Democratic Kampuchea.[59] Children and adults were required to attend nightly meetings—after long days of hard labor—and to announce their short-comings for all to hear. Throughout the sessions, Khmer Rouge leaders would admonish the people that "If you committed an error, criticize yourself first, then punish yourself." The people were to "Learn from model comrades, learn from our magnificent people, and forever learn from valuable documents, concentrate your mind to increase [your understanding of] the discipline of *Angkar* and the theories of Marx-Lenin, in order to raise your mental competency."[60] The purpose was to encourage public denunciations of malingerers, to expose "enemies" of the revolution, and to confess to deviant behavior. Locard continues:

> Accusations, arising almost always from commune leaders, invariably led to arrest, and then to liquidation. How often in public did people confess their wrongs, errors, flaws, or denounce their neighbors, almost always under threat and in fear for their own

55 H. Strub, "The Theory of Panoptical Control: Bentham's Panopticon and Orwell's Nineteen Eighty-Four," *The Journal of the History of the Behavioral Sciences* 25(1989): 40–49; at 42.

56 Foucault, *Discipline and Punish*, 48–49.

57 Foucault, *Discipline and Punish*, 33. See also James A. Tyner, *The Business of War: Workers, Warriors and Hostages in Occupied Iraq* (Aldershot, UK: Ashgate, 2006), 123–131.

58 Foucault, *Discipline and Punish*, 34.

59 Locard, *Pol Pot's Little Red Book*, 91.

60 Locard, *Pol Pot's Little Red Book*, 92–93.

safety? People did their best to be invisible, lower their heads, and keep their opinions to themselves.[61]

But these sessions could also be the site (and sight) of public torture and execution. Roeun Sam, for example, describes how the Khmer Rouge assembled a group of children at a place called Thunder Hill. The meeting took place at a temple, children sat in front. Two prisoners were called before the children. A Khmer Rouge official informed the children that "If someone betrays Angka, they will be executed. We want everyone to know that these people are bad examples, and we don't want other people to be like this." Sam explains that the Khmer Rouge forced the children to sit in front of the two prisoners; the children were warned, "If anyone cries or shows empathy or compassion for this person, they will be punished by receiving the same treatment."[62] Sam describes what happened next:

> Angka told someone to get the prisoner on his knees. The prisoner had to confess what he had done wrong. Then the prisoner began to talk but he didn't confess anything. Instead, he screamed, 'God, I did not do anything wrong. Why are they doing this to me? I work day and night, never complain, and even though I get sick and I have a hard time getting around, I satisfy you so you won't kill people. I never thought to betray Angka. This is injustice ... ' Suddenly one of [the Khmer Rouge guards] hit him from the back, pushed him, and he fell face to the ground. It was raining. We sat in the rain, and then the rain became blood. He was hit with a shovel and then he went unconscious and began to have a seizure. Then Angka took out a sharp knife and cut the man from his breastbone all the way down to his stomach. They took out his organs ... They tied the organs with wire on the handlebars of a bicycle and biked away, leaving a bloody trail.[63]

This brutal performance was intended to instill fear and ambiguity among the children. They were left bewildered, scared, confused. Why had he been killed? What did he do wrong? Afterwards, Sam explains: "I was now a prisoner in my mind and my body. My mind says, Don't remember, because this could be me. The air smelled like blood. Clear rain drops coming from the sky became blood ... Then Angka told us to get in line, and we all headed back to the place we lived."[64]

Another common form of public torture and execution occurred within both formal and make-shift prisons. Here, the audience were other prisoners, who were forced to witness the brutal killing of other people. Haing Ngor describes his witnessing the murder of a pregnant women while being held prisoner and tortured:

> Just before they put the plastic bag over my head, I glimpsed the pregnant lady next to me. She already had a bag over her head and she was kicking convulsively with both feet ... [Later they] took the bag off the pregnant lady next to me, but it was too late. She had died of suffocation. Then he [the guard] ripped her blouse apart and pulled down her sarong. Then he picked up his rifle, which had a bayonet attached. He pushed her legs apart and

61 Locard, *Pol Pot's Little Red Book*, 91.

62 Sam, "Living in Darkness," 76.

63 Sam, "Living in Darkness," 76.

64 Sam, "Living in Darkness," 77. Sam continues that she later felt angry and sorrowful; she was disturbed that she was unable to help the prisoner.

jammed the bayonet into her vagina and tried to rip upward but the pubic bone stopped the blade so he pulled the bayonet out and slashed her belly from her sternum down below her navel. He took the fetus out, tied a string around its neck and threw it in a pile with the fetuses from the other pregnant women. Then he reached into her intestines, cut out her liver, and finally sliced her breasts off with a sawing motion of his blade ... The flies whooshed around the body of the poor woman, whose crime had been marrying a Lon Nol soldier.[65]

In his account, Ngor does not attempt to explain the seemingly inexplicable. He simply notes of the Khmer Rouge, "It was nothing for them to cut someone open. Just a whim."[66] From a Foucauldian stand-point, however, the horrific murder and disemboweling of the woman is highly symbolic. It is not the individual bodies—those living people with names, families, and histories—but rather what their bodies signify in the moment of death *in the minds of the killers*. For the Khmer Rouge guard, the actual "identity" of the pregnant woman was inconsequential. What mattered was that she represented, in his mind, the "enemy," a "traitor" to *Angkar*. The torture and killing of the woman was an act of purification, an act to eliminate a threatening social group from the imagined society being constructed.

Haing Ngor's description of the murdered pregnant woman also highlights the often sexual symbolism that characterizes torture and execution. Historically, war-related rape and other forms of sexual violence have been (mis)characterized and dismissed as a private crime, or the result of an unfortunate few soldiers during time of war.[67] However, rape and other gendered-specific forms of violence should be viewed as deliberate instruments of discipline and violence. These are techniques of terror that inflict both physical and psychological trauma. These techniques also carry crucial symbolic components, derived in part from the so-called "natural" role of women to bear children.[68] According to Eisenstein, bodily violation—including rape and genital mutilation—destroys established gendered stereotypes. In particular, a "violated female is no longer a woman that a man wishes to lay claim to. In war [or genocide] rape, females are reduced to their patriarchal definition as a body vessel and also denied the status of privileged womanhood. In war rape the woman is totally occupied ... "[69] Consequently, women's bodies are battlegrounds; gendered forms of violence are techniques that degrade not just individual women, but to strip humanity from a larger group.

Ironically, in Democratic Kampuchea, cadre were instructed to *not* rape women. Sexual activity—outside the act of sex for reproduction—was discouraged by the

---

65 Ngor, *Survival in the Killing Fields*, 266.

66 Ngor, *Survival in the Killing Fields*, 266.

67 Binaifer Nowrojee, *Shattered Lives: Sexual Violence during the Rwandan Genocide and its Aftermath* (New York: Human Rights Watch, 1996), Patricia H. Hynes (2004) "On the battlefield of women's bodies: An overview of the harm of war to women," *Women's Studies International Forum* 27(2004): 431–445.

68 Nira Yuval-Davis, "Women and the biological reproduction of 'the nation.' *Women's Studies International Forum* 19(1996): 17–24; at 17.

69 Zillah Eisenstein, *Sexual Decoys: Gender, Race and War* (New York: Zed Books, 2006), 28.

Khmer Rouge. To be sure, these lessons were not always followed, especially among the senior leadership.[70] Nevertheless, numerous individuals found themselves arrested, tortured, and sometimes executed for sexual violations, including "adultery" between consenting adults.[71] What, then, are we to make of the gruesome sexual violation, degradation, and murder of the pregnant woman? Sadly, her's was not an isolated case. According to Locard, the mutilation of women's breasts and sexual organs was probably common, at the moment of execution.[72]

Rape, sexual violence, and gendered forms of mutilation are techniques that serve to both discipline and regulate bodies. These techniques—and the deliberate public "performance" of these acts—are used to perpetuate a morally exclusive geographical imagination of territorial integrity. These practices, through a focus on the reproductive system and of specific bodily parts, serve to destroy the future capacity of both individual bodies—even if the victim lives—and the collective population from reproduction. These acts, targeted at the reproduction of society function as exercises of power: symbolic attempts on behalf of the killers to construct their society anew.

These gendered acts, lastly, conform with the Khmer concept of disproportionate revenge. As Hinton explains, revenge within traditional Cambodian society is premised on a logic of debt and reciprocal exchange. The idea of disproportionate revenge suggests that when a person seeks vengeance, he or she may not only "defeat" the offender but also elevate's one's own honor. However, in the process one must also "completely defeat the enemy" to deter further retaliation. Hinton writes that "those who bear a grudge know that after they have exacted revenge, their adversary will in turn desire to repay the bad deed. To prevent the cycle from continuing, it may be in the avenger's interest to make a preemptive strike that will mute this desire by fear or death." The most extreme form of this type of revenge, according to Hinton, consists of killing one's enemies and their families.[73] Through the performance of disproportionate revenge, the Khmer Rouge sought to destroy entirely families and thereby provide a blank space to be inhabited. The specific targeting of (pregnant) women and their fetuses becomes a means of eliminating future generations of impure people. This also accounts for the crime of the pregnant woman witnessed by Haing Ngor: She was married to an enemy soldier and thus, her reproductive ability was viewed as doubly threatening.[74]

The public display of torture and execution thus served to reify the authority of Angkar. This is succinctly described by Ngor: "The soldiers took captives morning and afternoon. Instead of marching them away immediately, the soldiers made public examples of them, tying them to trees and shouting to anyone who would listen

---

70 Cases of rape, both individually and collectively, have been reported. See Locard, *Pol Pot's Little Red Book*, 258.

71 Locard, *Pol Pot's Little Red Book*, 257.

72 Locard, *Pol Pot's Little Red Book*, 258.

73 Alexander L. Hinton, *Why Did They Kill? Cambodia in the Shadow of Genocide* (Berkeley: University of California Press, 2005), 68–70.

74 These practices in fact have a long history in Cambodian society. Prior to the nineteenth-century, for example, Cambodian kings would, after winning a war, kill not only their subdued foes, but also their entire family lines. See Hinton, *Why Did They Kill?*, 71.

what they had done wrong."[75] These practices constituted a physical manifestation of spatial purification. However, not all acts of violence were visible. Many, in fact, took place in secrecy. Numerous people simply disappeared. Some were taken to torture facilities, such as Tuol Sleng, or less institutionalized settings. These sites were, in the words of Roeun Sam, the places the Khmer Rouge took people to kill. Ouk Villa, for example, describes waking up one night and, as a curious child, running off at night. "I ran through the rice fields and the bushes ... Suddenly I saw three men who were tied up and being led by a militiaman to another small bush. From a distance I saw the militiaman force the men to kneel down on the edge of a big pit. A minute later the men were clubbed to death with a hoe. I could see this clearly in the moonlit night."[76]

Prak likewise explains how the Khmer Rouge would gather people together, take them to an isolated spot, and kill them. He writes, "The people had thought they were going somewhere to work, but none came back."[77] These secret tortures were not to publically announce the supremacy of *Angkar*, but were used as practices of discipline. The sudden—and random—disappearance of other bodies, the not knowing of their fate: these instilled a terror that permeated daily life. As Ngor writes, "People carried within them an unspoken fear. They worried about their own survival, and they didn't trust anyone else, even their spouses."[78] Indeed, based on survivor's accounts, the most effective technique of discipline was the randomness of violence and the ever present fear of death. Sarom Prak describes, for example, that "Some people who accidentally broke the knives, hoes, axes, and plows they were working with were slaughtered by the Khmer Rouge."[79] Teeda Butt Mam remembers that "A man would be killed if he lost an ox he was assigned to tend. A woman would be killed if she was too tired to work. Human life wasn't even worth a bullet. They clubbed the back of our necks and pushed us down to smother us and let us die in a deep hole with hundreds of other bodies."[80]

Sreytouch Svay-Ryser was only seven years-old when the Khmer Rouge came to power. Forced to leave Phnom Penh and move to a collective in Battambang, Svay-Ryser recalls the constant fear that *Angkar* would discover that her brother had previously served in the Lon Nol army. She writes, "We knew that any time or at any hour *Angkar* could come to get him. *Angkar* usually came and took people away during the night. So when it got dark, we couldn't sleep."[81] And Roeun Sam, the 14 year-old girl introduced earlier, describes how this form of terror and fear was induced under the Khmer Rouge:

---

75 Ngor, *Survival in the Killing Fields*, 277.

76 Villa, "A bitter life," 116.

77 Prak, "The Unfortunate Cambodia," 69.

78 Ngor, *Survival in the Killing Fields*, 300.

79 Sarom Prak, "The unfortunate Cambodia," in *Children of Cambodia's Killing Fields: Memoirs by Survivors*, compiled by Dith Pran and edited by Kim DePaul (New Haven, CT: Yale University Press, 1997), 67–71; at 70.

80 Mam, "Worms from our skin," 12.

81 Sreytouch Svay-Ryser, "New Year's Surprise," in *Children of Cambodia's Killing Fields: Memoirs by Survivors*, compiled by Dith Pran and edited by Kim DePaul (New Haven, CT: Yale University Press, 1997), 35–41; at 38.

When night came I always worried. I stayed up even when they told us to go to sleep. Angkar walked around with a flashlight at night to see who was asleep and who wasn't. I was afraid that maybe next time it would be me. I would die before I saw the sun rise. I had little rest, and then I heard the whistle and inside I sighed, 'Oh, I'm alive!' I got up and got in line. From one night to the next it was the same.[82]

As Ngor concludes, "What was worse than hunger was the terror, because we couldn't do anything about it. The terror was always there, deep in our hearts. In the late afternoon, wondering whether the soldiers would choose us as their victims. And then feeling guilty when the soldiers took someone else. At night, blowing out our tiny oil lanterns so the soldiers wouldn't notice the light and come investigate, and then lying awake and wondering whether we would see the dawn. Waking up the next day and wondering whether it would be our last."[83]

In their quest to promote complete *conformity*, the consequences of *non*conformity were well understood by the Khmer Rouge. Through brutal systems of terror, the Khmer Rouge attempted to fix the spatiality of Democratic Kampuchea as a mimetic representation of their own geographical imaginations. Any resistance, any self-identity, any contestation over the subject and the state, were to be eliminated. This was accomplished through torture, starvation, and execution. Once defined, those who found themselves marginalized as the "Other" were subject to extreme practices of material deprivation and extermination. Jackson, indeed, refers to the Cambodian revolution as a "revolution by eradication."[84] Conceptually, we understand the eradication of people as "ethnic cleansing." Ideologically, the Khmer Rouge sought to construct a new, pure society not through transforming past geographies, but by erasing those spaces. The mass violence unleashed by the Khmer Rouge was not, in this sense, irrational or insane. Rather, it reflects, in Hannah Arendt's words, the "banality of evil." The violence meted by the Khmer Rouge was deliberate. And from the perspective of the Khmer Rouge, violence was functional.

Within the geographical imagination of the Khmer Rouge, the killing and death of bodies had a clear and distinct purpose: a systematic eradication of bodies that did not conform with the imagined geographies of Democratic Kampuchea. Whether the people died via starvation, lack of medical care, or execution was inconsequential. Haing Ngor, for example, explains that these Othered people, to the Khmer Rouge, "weren't quite people. [They] were lower forms of life, because [they] were enemies. Killing [them] was like swatting flies, a way to get rid of undesirables."[85]

Bodies, within Democratic Kampuchea, were to serve *Angkar* and the revolution. Bodies were thus viewed pragmatically. Life or death decisions hinged on whether any given body was perceived as having "use-value" for the Khmer Rouge. Indeed, the Khmer Rouge's oft-used saying "If you live there is no gain. If you die there is no loss," approaches this conception of life and death. The phrase speaks to the

82 Sam, "Living in Darkness," 78.
83 Ngor, *Survival in the Killing Fields*, 311.
84 Karl D. Jackson, "The Ideology of Total Revolution," in *Cambodia, 1975–1978: Rendezvous with Death*, Karl D. Jackson (ed.) (Princeton: Princeton University Press, 1989), 37–78; at 56.
85 Ngor, *Survival in the Killing Fields*, 247.

liminal position of bodies *and* spaces within Democratic Kampuchea. All bodies were inconsequential on their own. Their regime, in this sense, was totalizing; the persecution of people was not limited to "racial" groupings, as in other "genocides." Rather, the Khmer Rouge imagined a complete transformation of society whereupon all inhabitants were subjected to disciplinary techniques.

Hinton relates a particularly salient example of how the Khmer Rouge viewed their attempts to construct the *place* of Democratic Kampuchea. According to Hinton, one Khmer Rouge official likened the process of state-building to that of separating "spoiled" from "unspoiled" fruit or of separating wheat from chaff. When one approaches a basket of fruit, one could, for example, pick out the bad pieces, leaving the other pieces untouched. Conversely, one could overturn the *entire* basket and only put back the good pieces. Hinton suggests that this metaphor directly references the "othering" process that epitomizes Khmer Rouge practice. Certain groups were to be excluded (i.e., not placed back into the DK basket) from a newly "overturned" social order and were labeled as dirty and impure (i.e., spoiled fruit that should be discarded).[86] Bodies that have been spoiled cannot be returned to a pure, untainted state; they must be physically eliminated—destroyed—from the larger social body. In Democratic Kampuchea, the *entire* apple basket was to be overturned in the elimination of spoiled pieces.

The Khmer Rouge sought to *normalize* Democratic Kampuchea. As Foucault explains, "When you have a normalizing society, you have a power which [facilitates] an indispensable precondition that allows someone to be killed, that allows others to be killed."[87] And within Democratic Kampuchea, many "types" of bodies were conceived as expendable: the Vietnamese, the Buddhist clergy, former Lon Nol soldiers and government officials. Even the sick and the disabled were often conceived as expendable. Locard discusses how the Khmer Rouge viewed the "sick" and the "disabled" in Democratic Kampuchea.[88] According to Locard, "the sick could not be anything other than malingerers, because they could or would not work, and therefore, were sabotaging the revolution."[89] The following slogans capture these sentiments: "We absolutely must remove [from society] the lazy; it is useless to keep them, else they will cause trouble. We have to send them to hell;" "The sick are victims of their own imagination;" and "We must wipe out those who imagine they are ill, and expel them from our society!" Locard explains that all Cambodians "heard these rebukes shouted at them with violence and unusual harshness every time they had a fever or fell ill."[90]

---

86 Hinton, 2002, 81. This process is evident also in the Khmer Rouge slogan, "The winnowing basket separates the wheat from the chaff," quoted in Locard, *Pol Pot's Little Red Book*, 182.

87 Michel Foucault, *'Society Must be Defended': Lectures at the Collège de France, 1975–76*, Arnold I. Davidson (ed.), translated by David Macey (New York: Picador, 2003), 256.

88 Locard, *Pol Pot's Little Red Book*.

89 Locard, *Pol Pot's Little Red Book*, 187–188.

90 Locard, *Pol Pot's Little Red Book*, 188. Locard also explains that the reference to "imagination" is somewhat ambiguous. In Khmer Rouge parlance, the term could also mean "ideological frame of mind." Consequently, those accused of malingering were behaving so because their minds were infected with the ideology of the old society; they were not pure of

Not all "spoiled" people, however, were immediately subject to torture and execution. Even those who were eventually to be discarded could, from the stand-point of the Khmer Rouge, serve a productive function for Democratic Kampuchea.

The Body as a Productive Force

The Khmer Rouge established a totalitarian system that was predicated on terror in order to maintain discipline. Consequent was a regimented, predictable, hyper-orderly society, one that negated human will, spontaneity and creativity. Discipline via corporeal control was to produce total conformity. And within the totalitarian spaces of Democratic Kampuchea, all facets of humanity were monitored and disciplined. Bodies were to exist solely for *Angkar*, the state, and the revolution. Bodies, in short, were to be either economically and/or politically useful.

Foucault explains that the "body" "is bound up, in accordance with complex reciprocal relations, with its economic use; it is largely as a force of production that the body is invested with relations of power and domination; but, on the other hand, its constitution as labour power is possible only if it is caught up in a system of subjection ...; the body becomes a useful force only if it is both a productive body and a subjected body."[91] Furthermore, as Foucault elaborates, "Discipline increases the forces of the body (in economic terms of utility) and diminishes these same forces (in political terms of obedience)." He continues that discipline "dissociates power from the body; on the one hand, it turns it into an "aptitude," a "capacity," which it seeks to increase; on the other hand, it reverses the course of the energy, the power that might result from it, and turns it into a relation of strict subjection."[92]

It is necessary, therefore, to consider more closely how the Khmer Rouge conceived of the body as a productive force. Such a consideration, moreover, must address the twin concepts of authority and sovereignty. I begin with the proposition that the principal expression of sovereignty resides, to a large degree, in the power and the capacity to dictate who may live and who must die.[93] This point is elaborated by Foucault.[94] In the classical theory of sovereignty the right of life and death was one of sovereignty's basic attributes. In other words, to say that "the sovereign has a right of life and death means that he [sic] can ... either have people put to death or let them live."[95] Foucault, furthermore, suggests that the sovereign cannot grant life in the same way that he or she can inflict death. The right of life and death "is always exercised in an unbalanced way: the balance is always tipped in favor of death." Consequently, the "very essence of the right of life and death is actually the right to

---

mind or of heart. This is made clear in the following slogan: "If you have the disease of the old society, take a dose of Lenin as medication." See Locard, *Pol Pot's Little Red Book*, 188.

91 Foucault, *Discipline and Punish*, 25–26.

92 Foucault, *Discipline and Punish*, 138.

93 Achille Mbembe, "Necropolitics," *Public Culture* 15(2003): 11–40; at 11.

94 Michel Foucault, "Technologies of the Self," in *Ethics: Subjectivity and Truth, Essential Works of Foucault, 1954–1984*, P. Rabinow (ed.) (New York: The New Press, 2000), 223–251.

95 Foucault, "Technologies of the Self," 240.

kill: it is at the moment when the sovereign can kill that he [sic] exercises his right over life."[96] Life and death, therefore, are removed from the realm of the "natural" and thus fall within the field of governance. Life and death become *political*.

Derived from a reading of French history,[97] Foucault makes the argument that this conception of sovereignty shifted with the emergence of non-disciplinary practices and the transition from a concern with the anatomo-politics of the body to that of biopolitics. Foucault identified a shift that occurred between the seventeenth and eighteenth centuries, a shift that would profoundly affect the machinations of governance. It was during this period that the first steps were taken toward a "scientific" study of population.[98] As manifest in the writings of John Graunt, William Petty, William Farr, and Thomas Malthus, governments began to reorient their concerns away from *bodies* and toward *populations*.

Through the advents of censuses and surveys, statistical rates and ratios, a scientific approach to people was developed. And these knowledges were to serve the state. As Legg writes, "Here was a government that became obsessed with statistics concerning

---

96 Foucault, "Technologies of the Self," 240.

97 Throughout this section I draw heavily on Foucault's conceptualization of sovereignty, based on his extensive studies of French history. To some, this may seem misplaced, that I should rely more so on traditional Khmer cultural understandings of power, authority, and sovereignty. However, I maintain that such a "Eurocentric" understanding of Khmer Rouge ideology is not so far-fetched as one might imagine. Recall that many of the leading ideologues of the Khmer Rouge were schooled in French institutions, and that many of the leading theorists—Khieu Samphan, Ieng Sary, and Pol Pot—received advanced educations in France. These individuals were well-schooled in (or, at least in Pol Pot's case, exposed to) *French* and *European* politics and governance. These men were trained in the French language and studied French history. Indeed, according to David Chandler, one of Pol Pot's most revered philosophers of Jean-Jacques Rosseau. In short, one must recognize that the Khmer Rouge's understanding of revolution and state sovereignty was derived more from European examples than from indigenous Khmer understandings. Concurrently, however, we should not lose sight that many practices of the Khmer Rouge did, in fact, resonate with traditional Khmer culture. For example, Khmer Rouge attitudes to violence *do* conform with Khmer concepts of sovereignty. In Cambodian folk-tales, for example, the act of murder is not normally conceived as an evil or as a uniquely reprehensible act. Rather, murder may assume a revelatory function, either in the sense that it serves as a prelude to a rebirth, or in that it triggers an act of salvation. Consequently, the killings of (conceived) corrupt and irredeemable elements was understood by the Khmer Rouge as a prelude to the birth of a moral and properly ordered society.

98 See Mitchell Dean, *Governmentality: Power and Rule in Modern Society* (Thousand Oaks, CA: Sage Publications, 1999); John Caldwell, "Demographers and the study of mortality: Scope, perspectives, and theory," *Annals of the New York Academy of Sciences* 954(2001): 19–34; Chris Philo, "Accumulating populations: Bodies, institutions and space," *International Journal of Population Geography* 7(2001): 473–490; Bruce Curtis, "Foucault on governmentality and population: The impossible discovery," *Canadian Journal of Sociology* 27(2002): 505–533; Chris Philo, "Sex, life, death, geography: Fragmentary remarks inspired by 'Foucault's population geographies,'" *Population, Space and Place* 11(2005): 325–333.

birth rates, morbidity and endemics, supplanting the earlier fear of epidemics."[99] Now, a focus on the body revealed it as being imbued with the mechanics of life and serving as the basis of various biological processes: propagation, births and mortality, the level of health, life expectancy and longevity.[100] There emerged, consequently, a concept of "populations," a large body of people constituting some kind of definable unit to which measurements could be applied.[101]

The emergence of "population" was not happenstance. Rather, the birth of this concept was a gradual process, both technical and theoretical, involving the development of statistics and census taking, and the flourishing of epidemiology, demography, and political philosophies as disciplines of knowledge.[102] The emergent field of demography, for example, was inextricably tied to the emergence of merchant capitalism in post-feudalistic Europe. Mercantilism is premised on the idea that a Nation's wealth is determined by the amount of precious metals it had in possession. Colonialism, consequently, was fundamental to the rise of merchant capitalism: colonies provided instant wealth, access to resources, cheap labor in the form of indentured servitude and slavery, and ready-made markets.[103]

Mercantilism was a self-propelling growth system in which the continued expansion of trade was deemed vital. Without growth, neither merchants nor those dependent on their trade (i.e., financiers) could maintain their position. Consequently, mercantilist writers sought to encourage population growth by various practices, including penalties for non-marriage, encouragements to get married, lessening penalties for illegitimate births, and promoting immigration of productive laborers.[104] Concurrently, the ascension of mercantilism and, later, industrial capitalism, contributed to the widespread practice of insurance and life annuities, banking and loan systems, credit transfers and shares in stock, and speculation in commodity futures.[105]

The emergence of demography coincided with the development of liberalism as a political economic philosophy. According to Dean, "the point at which population ceases to be the sum of the inhabitants within a territory and becomes a reality *sui generis* with its own forces and tendencies is the point at which this dispositional government of the state begins to meet a government through social, economic, and biological processes."[106] In other words, "population" becomes an object in its own right. From this point forward, it was possible to speak of a state's population, as if that population had a transcendental existence and experience above and beyond the

---

99 Stephen Legg, "Foucault's Population Geographies: Classifications, Biopolitics and Governmental Spaces," *Population, Space and Place* 11(2005): 137–156; at 141.

100 Michel Foucault, *The History of Sexuality, Volume 1: An Introduction*, translated by Robert Hurley. (New York: Vintage Books, 1990), 139.

101 Caldwell, "Demographers," 20.

102 Dean, *Governmentality*, 108.

103 John R. Weeks, *Population: An Introduction to Concepts and Issues*, 9th edition (Belmont, CA: Wadsworth/Thomson Learning, 2005), 73.

104 Weeks, *Population*, 73. See also Paul Knox, John Agnew, and Linda McCarthy, *The Geography of the World Economy*, 4th edition (London: Arnold, 2003), 127.

105 Knox et al., *Geography of the World Economy*, 130–131.

106 Dean, *Governmentality*, 95–96.

government. Moreover, it is here that we can locate the construction of populations into sub-groups that contribute to or retard the general welfare and life of the population as a whole. For Dean, it is this proclivity that led to the discovery among the population *writ large* of criminal classes, for example, or the feeble-minded and the imbeciles, the inverts and degenerates, the unemployable and the abnormal.[107]

It is the notion of a "population," according to Dean, that makes possible the elaboration of a distinctly liberal government. Governments required precise, objective information: knowledges of its population and its resources. Elaborate and extensive administrative apparatuses were developed to facilitate this information and to translate it into actual policies and practices. Indeed, as Dean continues, the concept of "population" introduced several key elements that significantly influenced the art of governance. For one, the idea of population introduced a different conception of the governed. The members of a population were conceived not as subjects bound together in a territory, and who were obliged to submit to a sovereign. Rather, populations were conceived as living and working social beings, with their own customs, habits, and histories. Second, populations were defined in relation to matters of life and death, health and illness, propagation and longevity, all of which could be *known* and *manipulated* by statistical and demographic instruments. Third, the concept of population imparted an idea of a collective, the knowledge of which was irreducible to the knowledge that any of a population's members might have of themselves.[108] This was a crucial ontological shift for, as Caldwell explains, those scholars who came to identify themselves as demographers were suspicious of the study of individuals and small groups; they believed that such bodies were significant only when it could be shown what fraction of a larger population they constituted. Demographers, consequently, looked for regularities in populations or sub-populations, as well as for contrasts between sub-populations.[109]

Given such an ontological position, Dean explains that now one could ask:

> about the historical development of certain aspects of the population: its marriage customs, the number of marriages that are usually conjoined, at what ages, how many children are produced by these marriages, the customs that take place within the family, the price of labour and its variation, and the happiness of the working population at a given time. In short, we can know a population, and its industry, customs and history, as a collective identity that is not constituted by political or governmental institutions or frameworks.[110]

Supervision of these population processes was effected through a series of interventions that Foucault terms "biopolitics." Biopolitics entail "a set of processes such as the ratio of births to deaths, the rate of reproduction, the fertility of a population, and so on. It is these processes ... together with a whole series of related economic and political problems ... [that] become biopolitics' first objects of knowledge and the targets it seeks to control."[111] Biopolitics, therefore, addresses

---

107 Dean, *Governmentality*, 107.
108 Dean, *Governmentality*, 107.
109 Caldwell, "Demographers," 21.
110 Dean, *Governmentality*, 107.
111 Foucault, *Society Must be Defended*, 243.

the population, with population as a political-economic problem. New technologies, such as population forecasts, were not applied to individual bodies, but rather to collectives—populations. And the purpose of these techniques, according to Foucault, was not to modify any given phenomenon as such, or to modify a given body insofar as he or she is an individual, but instead to intervene at the level at which these general phenomena are determined.[112] Individual deaths, for example, matter less than the overall mortality rate of the state.

Foucault suggests that as the state developed, power was decreasingly viewed as the right to take life and increasingly seen as the right to intervene to make live. Such a right was made possible, for example, through the elimination of random elements of accidents and so forth. Power, consequently, was exercised not over death, but over mortality. Once this transition occurred, Foucault asks, "How is it possible for a political power to kill, to call for deaths, to demand deaths, to give the order to kill, and to expose not only its enemies but its own citizens to the risk of death? Given that this power's objective is essentially to make live, how can it let die?"[113]

The answer, for Foucault, lies in a concept he terms "State racism," for it is at this point that racism (or any other classification scheme) intervenes to separate bodies into populations: populations that are categorized as "us" or "them," "good" or "bad," "friend" or "enemy." Foucault states bluntly: "It is primarily a way of introducing a break into the domain of life that is under power's control: the break between what must live and what must die."[114] It is a way of separating out types of bodies, of constructing different populations that are evaluated in such a way as to justify death over life. Foucault continues:

> The appearance within the biological continuum of the human race of races, the distinction among races, the hierarchy of races, the fact that certain races are described as good and that others, in contrast, are described as inferior: all this is a way of fragmenting the field of the biological that power controls. It is a way of separating out the groups that exist within a population. It is, in short, a way of establishing a biological type caesura within a population that appears to be a biological domain.[115]

The re-grouping of bodies into populations and sub-populations is a decidedly spatial process. It is an activity of dividing and separating people for political and economic purposes. However, there is an additional component to these socio-spatial separations, one that is immediately connected to questions of sovereignty and social justice. Foucault notes that "The fact that the other dies does not mean simply that I live in the sense that his death guarantees my safety; the death of the other, the death of the bad race, of the inferior race (or the degenerate, or the abnormal) is something that will make life in general healthier: healthier and purer."[116]

Following Foucault, we arrive at a practice of spatial purification in the context of state-building. The construction of a pure society—written onto pages erased clean

---

112 Foucault, *Society Must be Defended*, 246.
113 Foucault, *Society Must be Defended*, 254.
114 Foucault, *Society Must be Defended*, 254.
115 Foucault, *Society Must be Defended*, 254–255.
116 Foucault, *Society Must be Defended*, 255.

from all previous societies—requires the eradication and elimination of essentialized Others. These bodies, conceived as forever separate, could not (or rarely ever), in the geographical imagination of the Khmer Rouge, be re-educated, re-habilitated, or re-formed. There emerged a perceived necessity to simply exterminate the unwanted, impure bodies—if they were not used productively. As Ponchaud explains, the purge of the Cambodian population "was, above all, the translation into action of a particular vision of man [sic]: a person who has been spoiled by a corrupt regime cannot be reformed, he must be physically eliminated from the brotherhood of the pure."[117] This brutal vision is aptly illustrated by the various slogans disseminated over the radio and at communal meetings during the Khmer Rouge regime: "What is infected must be cut out" and "What is rotten must be removed."[118] Tens of thousands of deaths were ordered; but many more—hundreds of thousands—died not through direct orders, but because of particular policies enacted by the Khmer Rouge. These deaths were not necessarily intentional, but from the stand-point of the Khmer Rouge, the objective was (ostensibly) to "make live" for certain people. Those who did not conform died through malign neglect.

The horrific and constant killings and deaths metastasized into banality. As Haing Ngor describes: "For the Khmer Rouge in general ... the act of killing other human beings was routine. Just part of the job. Not even worth a second thought."[119] However, Ngor also notes that for many Khmer Rouge leaders—as opposed to the low-ranking soldiers—killing was more sophisticated. It was a political necessity. Ngor contends that "When [the Khmer Rouge leaders] talked about sacrificing everything for Angka, they meant it. Whatever got in the way of Angka's projects had to be eliminated, including people. To them, though, we weren't quite people. We were lower forms of life, because we were enemies."[120]

Conclusions

The project of the Khmer Rouge was a projection. It was an attempt to spatially and temporally promote a particular geographical vision of a modern, communist-based utopia. The spatial practices of purification sanctioned by Pol Pot and other high-ranking officials of the Khmer Rouge were designed to achieve a number of objectives: economic self-sufficiency of Democratic Kampuchea, centralization of authority within *Angkar*, the dissolution of family structures. A primary objective, however, and one that most directly affected the life and death of the people of Cambodia, was geographic. The systematic violence and the killing of (imagined) impure and irredeemable bodies were understood by the Khmer Rouge as a prelude to the construction of a moral and properly ordered society. The killing of Cambodia

---

117 Quoted in Quinn, "The Pattern and Scope of Violence," 185.

118 Quinn, "Pattern and Scope of Violence," 186.

119 Ngor, *Survival in the Killing Fields*, 246.

120 Ngor, *Survival in the Killing Fields*, 247. Ngor explains also that the lower-ranking soldiers killed because they were trained to kill; when ordered to take another's life, they obeyed automatically. See also Hinton, *Why Did They Kill?*

was not a by-product of Democratic Kampuchea: it was a deliberate and intentional political strategy.

Life had little value for the Khmer Rouge leadership beyond the purpose of serving *Angkar*, Democratic Kampuchea, and the revolution. This is pointedly expressed in the Khmer Rouge slogan: "To die is banal for the one who fights heroically."[121] All bodies, on their own, were presumed to be inconsequential. It was only within the context of the revolution, of the state of Democratic Kampuchea, that bodies acquired meaning and purpose. And for the Khmer Rouge, bodies were to be either productive workers (i.e., performing labor on collectives) and/or politically loyal (i.e., docile). Otherwise, bodies were considered as barriers to the culmination of the collective utopia imagined by the Khmer Rouge. This is why Pol Pot, on his death bed, continued to exhibit no remorse for the death he unleashed on Cambodia. From his point of view, the Khmer Rouge regime did not kill *anyone*; they simply removed the detritus of society that threatened the sovereignty of Democratic Kampuchea. And therein lies the most horrific aspect of the brutality of the Khmer Rouge: Their actions, as motivated by their geographical imaginations, were considered just.

---

121 Locard, *Pol Pot's Little Red Book*, 210.

Chapter 7

A Political Understanding of Genocide and Justice

"A revolution that does not produce a new space has not realized its full potential; indeed it has failed in that it has not changed life itself, but has merely changed ideological superstructures, institutions or political apparatuses. A social transformation, to be truly revolutionary in character must manifest a creative capacity in its effects on daily life, on language and on space."[1] Henri Lefebvre's thesis is foundational to the arguments I make in *The Killing of Cambodia*, specifically that the Khmer Rouge sought to erase the histories, geographies, and societies of Cambodia and, consequently, attempted to create a new state, that being Democratic Kampuchea. In the process, the Khmer Rouge operated within a particular corporeal ontology that—in their imaginations—legitimated and justified a series of brutal practices that led to the death of over two million people. From the perspective of Pol Pot and his followers, Cambodia, literally, had to be killed. And it was through this murder that Democratic Kampuchea was born.

From its origins as a disorganized and disparate social movement, the Communist Party of Kampuchea (CPK) emerged as the most violent and brutal apparatus of terror and murder since the Nazi Party held power in Germany. Between April 1975 and January 1979 the Khmer Rouge carried out a program of mass violence that is, in many respects, unparalleled in modern history. For three years, eight months, and 20 days, approximately two million people—including native Cambodians, Cambodian-born Vietnamese, and Sino-Khmers—died from starvation, inadequate medical care, torture, murder, and execution. The totaled number of deaths approaches one-third of Cambodia's pre-1975 population.

In *The Killing of Cambodia* I invert Lefebvre's notion of the "production" of space, namely through a concept of an "erasure" of space. For the Khmer Rouge leadership, it was not only necessary to write history anew, to begin with blank pages, but to write geography anew. The landscape of Democratic Kampuchea was not to reflect past geographies of sequent occupance. Instead, the former spaces of Cambodia needed to be smashed, as manifest in the destruction of urban areas and of human bodies.

The resultant "place" of Democratic Kampuchea was one of placelessness. In an attempt to construct a communist utopia, the Khmer Rouge produced a non-place, an alienated and inauthentic place. And who was to occupy this flatscape? The Khmer Rouge envisioned a new society, a collective that was stripped bare of individualism.

---

1 Henri Lefebvre, *The Production of Space*, translated by Donald Nicholson-Smith (Oxford, UK: Blackwell, 1991), 54.

The people of Cambodia were subjected to horrific disciplinary techniques of terror and murder. They were subject to forced labor and starvation. Individual bodies were inconsequential on their own; bodies were conceived as serving a purely pragmatic function: to serve *Angkar*, the revolution, and Democratic Kampuchea.

The killing of Cambodia was deliberate. It was justified in the geographical imagination of Pol Pot, Khieu Samphan, and the other visionaries of Cambodia's necropolis. And in this sense, the mass violence that stalked Cambodia was *genocidal in intent*. But my purpose in setting out on this project was never to argue for or against charges of genocide. While vitally important, I defer to the work of Youk Chhang, David Chandler, Ben Kiernan, Craig Etcheson, and Stephen Heder, among others. No, my goals are decidedly more modest.

In Chapter 1 I indicated that a pressing task is to uncover the "battle for truth." This task, following Roxanne Doty, is "not to reveal essential truths that have been obscured, but rather to examine *how* certain geographic representations underlie the production of knowledge and identifies and how these representations make various courses of action possible."[2] Now, as we near the conclusion of this book, we see that this understanding entails not simply a battle for truth, but rather *a battle for truth over the meaning of justice*.

Social justice often entails a conception of what is fair and unfair, and of the social arrangements necessary to ensure that members of society are treated justly.[3] This understanding is often reduced to concerns of distributive justice, of the moral distribution of benefits and burdens among society. David Smith explains that "A central issue in distributive justice is how to justify differential treatment, or how to identify the differences among people which are relevant to the particular attribute(s) to be distributed."[4] Rhetorically, the Khmer Rouge attempted to overcome such differential treatments through the elimination of difference. Ironically, the Khmer Rouge—discursively—understood Don Mitchell's argument that "both oppression and domination are exercised through difference; it is difference that is oppressed and it is differentially situated actors who dominate. Autonomy—the freedom to be who one is—requires not just the recognition of difference but also its social promotion."

In Democratic Kampuchea, differences were recognized—or, more properly, differences were constructed—and thus targeted for elimination. The Khmer Rouge claimed to create an egalitarian society through the annihilation of difference and, in so doing, they claimed to eliminate oppression and domination. But at the same time, autonomy was to be avoided—in the name of social justice. Their actions consequently are exceptionally important in our discussions of the universality or particularity of human rights.

---

2 Roxanne Lynn Doty, *Imperial Encounters: The Politics of Representation in North-South Relations* (Minneapolis: University of Minnesota Press, 1996), 5.

3 Caroline R. Nagel, "Social Justice, Self-Interest and Salman Rushdie: Reassessing Identity Politics in Multicultural Britain," in *Geography and Ethics: Journeys in a Moral Terrain*, James D. Proctor and David M. Smith (eds) (New York: Routledge, 1999), 132–146; at 133. Don Mitchell, *The Right to the City: Social Justice and the Fight for Public Space* (New York: The Guilford Press), 32.

4 David M. Smith, *Geography and Social Justice* (Cambridge, MA: Blackwell, 1994), 24.

David Harvey argues that to assert a situation is unjust presupposes that there are some universally agreed upon norms as to what we do or ought to mean by the concept of social justice.[5] Iris Young, however, counters this argument, noting that appeals to justice and claims of injustice are not a result; they are the starting point for a certain form of debate.[6] Who is to claim the existence of a set of pre-determined all-encompassing universal rights? Who decides? The Khmer Rouge made such an assertion. Indeed, if one were to ask the leaders of the CPK, they would no doubt couch their actions within the rhetoric of "just war." And were Pol Pot still alive, he would most likely describe the deaths that occurred in Democratic Kampuchea as "collateral damage" or "acceptable losses." We conclude, however, that the actions of the Khmer Rouge were not just; indeed, we conclude that their actions constitute a grave injustice. But, on what basis can we make this claim? Similar to the on-going debates over the meaning of genocide, we are left with the uncomfortable proposition that justice itself is discursive. It is subject to political and economic manipulation.

It is troubling to realize that the Khmer Rouge justified its actions and condoned the death of millions of its own citizens. The deaths, and the disciplinary techniques (confinement, relocation, torture) were justified and legitimated in terms of the promotion of a just and egalitarian society. The Khmer Rouge claimed to establish a universal understanding of right and wrong. And this is precisely the issue. Who ultimately makes the decision as to what constitutes right and wrong? In Cambodia, the Khmer Rouge claimed this right, and over two million people died.

In a further irony, Young claims that in debates over justice, people will formulate principles to support their claims.[7] This, the Khmer Rouge did. They extracted, albeit through torture, signed confessions to support their claims of righteousness. They operated through secretive state apparatuses, such as *Angkar*, to further justify their actions. Members of society were expected (forced) to internalize these sentiments— through nationalist songs, literature, slogans and schooling.

I may assert—and I do—that the practices of the Khmer Rouge constitute one of humanities worst injustices. But my assertion rests on equal footing as that of the Khmer Rouge. Is it possible to substantiate my claim while denying that of the Khmer Rouge? Or are we relegated, as Harvey might suggest, to a nihilistic state of passivity? Are we forever condemned to a relativistic morass whereby a progressive politics is impossible? I don't believe so.

Social justice is contextual and its meanings are constantly contested. In other words, claims to social justice are *political*. Claims to social justice are historically and geographically contingent and therefore cannot be based on appeals to universal laws. The quest for social justice is an ongoing *political process*. However, part-and-parcel of this political process, I suggest, is a greater engagement with morality and violence. Namely, if we are to condemn the Khmer Rouge for their actions,

---

5 David Harvey, *Justice, Nature and the Politics of Difference* (Oxford: Blackwell, 1996).

6 Iris M. Young, "Harvey's Complaint with Race and Gender Issues: A Critical Response," *Antipode* 30(1998): 36–42.

7 Young, "Harvey's Complaint."

we must condemn all acts of violence. We must foster and facilitate a practice of non-violence. Non-violence is not pacifism. Whereas pacifism is passive, non-violence is active. It is political. Mark Kurlansky explains that non-violence "is a means of persuasion, a technique for political activism, a recipe for prevailing." As a philosophy and political practice, the central belief of non-violence is political, that non violence is more effective than violence.[8] This must be prevalent in both our research and our teaching.

In order to work toward the elimination of violence and, especially, war, we must promote a different vision, a different geographical imagination of how the world *could* be, one based on peace and non violence. I concur with Harris and Morrison who argue that "If and when the desire for peace becomes strongly rooted in human consciousness, people will strive for it, demanding new social structures that reduce risks of violence."[9] One strategy, as academics, is thus to promote peace and non violence in our everyday activities. As Harris and Morrison conclude, "Human beings have a choice about how to live on this planet. Education, by influencing students' attitudes, information, and ideas about peace, can help create in human consciousness the moral strength that will be necessary to move toward a more peaceful future."[10]

Peace education is the promotion of non-militarized values.[11] It is the denial of violent practices as effective political techniques. Peace education, according to Lennart Vriens, "does not try to make peace by means of education; peace education cannot make peace directly, but instead aims to make people able to judge in the service of political co-responsibility for peace."[12] Educating for peace, therefore, signals an attempt to transform society by creating a non violent consciousness that condemns violent behavior. Professors, consequently, can teach about the problems of war and peace—for example, of how genocides and wars are *justified* as political processes. Peace education should be designed to recognize, challenge, and change the thinking that has supported oppressive societal structures; it should reveal conditions that trigger violence, ideological rivalries, and national policies that maintain arms races, military systems, and inequitable economic priorities; it should empower people with the skills and attitudes to create a non-violent world.[13]

---

8 Mark Kurlansky, *Nonviolence: Twenty-Five Lessons from the History of a Dangerous Idea* (New York: Modern Library, 2006), 6–7.

9 Ian M. Harris and Mary Lee Morrison, *Peace Education*, 2nd edition (London: McFarland Co., 2003), 15.

10 Harris and Morrison, *Peace Education*, 177.

11 On the promotion of militarism, see Chalmers Johnson, *The Sorrows of Empire: Militarism, Secrecy, and the End of the Republic* (New York: Metropolitan Books, 2004); Andrew J. Bacevich, *The New American Militarism: How Americans are Seduced by War* (Oxford: Oxford University Press, 2005); David Keen, *Endless War? Hidden Functions of the 'War on Terror'* (London: Pluto Press, 2006); and Ismael Hossein-Zadeh, *The Political Economy of US Militarism* (New York: Palgrave Macmillan, 2006).

12 Lennart Vriens, "Peace Education: Cooperative Building of a Humane Future," *Pastoral Care in Education* 15(1997): 25–30; at 27.

13 Susan Opotow, Janet Gerson, and Sarah Woodside, "From Moral Exclusion to Moral Inclusion: Theory for Teaching Peace," *Theory into Practice* 44(2005): 303–318; at 305.

Dith Pran, survivor of Cambodia's genocide, writes that "The dead are crying out for justice. Their voices must be heard. It is the responsibility of the survivors to speak out for those who are unable to speak, in order that the genocide and holocaust will never happen again in this world."[14] As academics, as researchers, as educators, it is our responsibility as well.

---

14 Dith Pran, "Compiler's Note," in *Children of Cambodia's Killing Fields: Memoirs by Survivors*, compiled by Dith Pran and edited by Kim DePaul (New Haven, CT: Yale University Press, 1997), ix–x; at x.

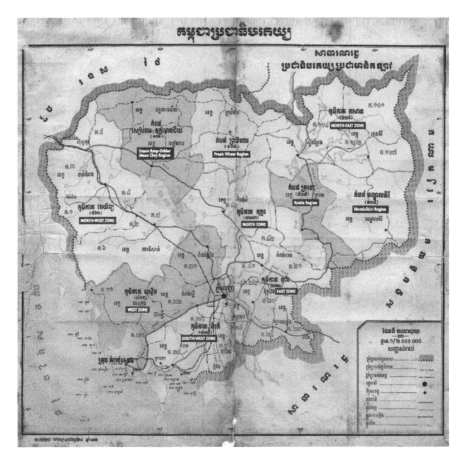

**Plate 1 This map was published by Democratic Kampuchea's
Ministry of Education in 1977**

Source: Photo courtesy of the Documentation Center of Cambodia.

Plate 2 CPK General Secretary Saloth Sar, alias Pol Pot

**Plate 3a On April 15, 1975 Khmer Rouges entered
          Cambodia's capital, Phnom Penh**
Source: Photo courtesy of the Documentation Center of Cambodia.

Plate 3b Khmer Rouge troops in Phnom Penh
Source: Photo courtesy of the Documentation Center of Cambodia.

Plate 3c Khmer Rouge troops in Phnom Penh
Source: Photo courtesy of the Documentation Center of Cambodia.

Plate 3d Khmer Rouge troops in Phnom Penh
Source: Photo courtesy of the Documentation Center of Cambodia.

Plate 3e Khmer Rouge troops in Phnom Penh
Source: Photo courtesy of the Documentation Center of Cambodia.

Plate 4 After the evacuation, Phnom Penh was a ghost town
Source: Photo courtesy of the Documentation Center of Cambodia.

Plate 5 Prior to building Democratic Kampuchea, the Khmer Rouge attempted to un-make Cambodia

Source: Photo courtesy of the Documentation Center of Cambodia.

Plate 6 Phnom Penh after the evacuation

Source: Photo courtesy of the Documentation Center of Cambodia.

Plate 7 Communal eating in dining hall. Throughout Democratic Kampuchea, a rigid conformity was imposed on all facets of daily life
Source: Photo courtesy of the Documentation Center of Cambodia.

Plate 8 Prison staff eating communally at Tuol Sleng. Kaing Buek Eav, alias Duch, is seen standing to the side
Source: Photo courtesy of the Documentation Center of Cambodia.

Plate 9 Two female Khmer Rouge cadre
Source: Photo courtesy of the Documentation Center of Cambodia.

Plate 10 The CPK literally rebuilt Cambodia. Here, women are seen working on irrigation canals in a Khmer Rouge labor camp
Source: Photo courtesy of the Documentation Center of Cambodia.

Plate 11 Cambodians transporting dirt on an irrigation project
Source: Photo courtesy of the Documentation Center of Cambodia.

Plate 12 Workers constructing a dam during Democratic Kampuchea
Source: Photo courtesy of the Documentation Center of Cambodia.

Plate 13 Workers constructing a dam
Source: Photo courtesy of the Documentation Center of Cambodia.

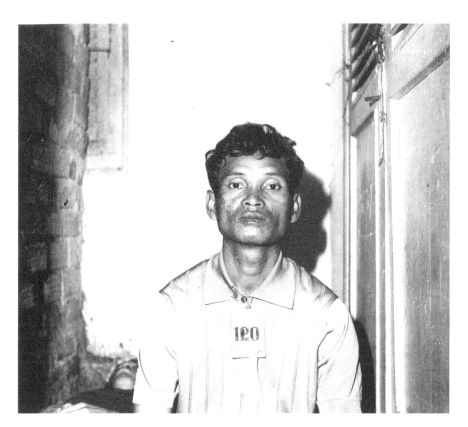

Plate 14 Prisoner, Tuol Sleng
Source: Photo courtesy of the Documentation Center of Cambodia.

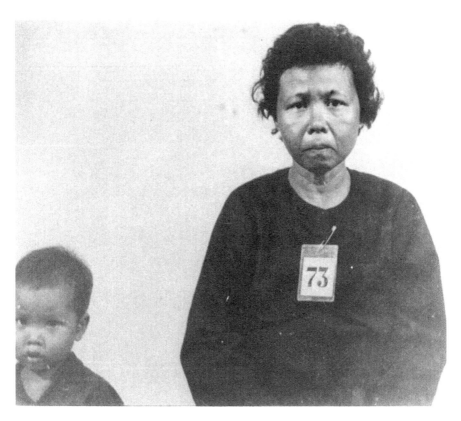

Plate 15 Prisoners, Tuol Sleng
Source: Photo courtesy of the Documentation Center of Cambodia.

Plate 16 Blood-stained interrogation room at Tuol Sleng
Source: Photo courtesy of the Documentation Center of Cambodia.

Plate 17 A prisoner following his suicide (with hand-written notes detailing the event)
Source: Photo courtesy of the Documentation Center of Cambodia.

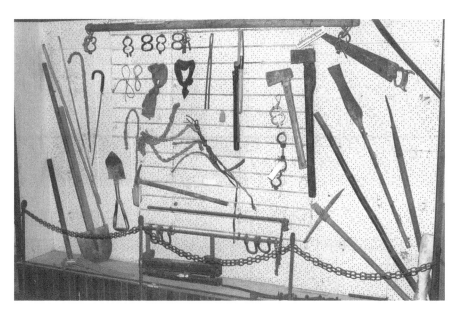

Plate 18 Tools used for torture and murder at Tuol Sleng
Source: Photo courtesy of the Documentation Center of Cambodia.

Plate 19 Mug shots of prisoners tortured and executed at Tuol Sleng
Source: Photo courtesy of the Documentation Center of Cambodia.

**Plate 20 The killing fields at Choeung Ek, a site of mass executions
        outside of Phnom Penh**
Source: Photo courtesy of the Documentation Center of Cambodia.

Plate 21 Mass grave, Choeung Ek
Source: Photo courtesy of the Documentation Center of Cambodia.

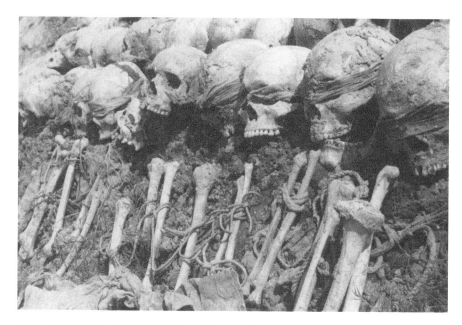

Plate 22 Skulls with blind-folds, Choeung Ek
Source: Photo courtesy of the Documentation Center of Cambodia.

Plate 23 The killing of Cambodia
Source: Photo courtesy of the Documentation Center of Cambodia.

Plate 24 Survivors of the killing fields returning home in 1979
Source: Photo courtesy of the Documentation Center of Cambodia.

Plate 25 Cambodians returning to an uncertain future
Source: Photo courtesy of the Documentation Center of Cambodia.

Plate 26 Refugees returning after the killing fields
Source: Photo courtesy of the Documentation Center of Cambodia.

Plate 27 Refugees returning after the killing fields
Source: Photo courtesy of the Documentation Center of Cambodia.

Plate 28 Refugees returning after the killing fields
Source: Photo courtesy of the Documentation Center of Cambodia.

Bibliography

Agnew, John, *Place and Politics* (Boston: Allen and Unwin, 1987).

—— "Space: Place," in *Spaces of Geographical Thought: Deconstructing Human Geography's Binaries*, Paul Cloke and Ron Johnston (eds) (Thousand Oaks, CA: Sage Publications, 2005), 81–96.

Anderson, Kay and Gale, Faye (eds), *Inventing Places: Studies in Cultural Geography* (London: Belhaven Press, 1992).

Ashcroft, Bill and Ahluwalia, *Edward Said* (New York: Routledge, 1999).

Augé, Marc, *Non-Places: Introduction to an Anthropology of Supermodernity* (London: Verso, 1995).

Bacevich, Andrew J., *The New American Militarism: How Americans are Seduced by War* (Oxford: Oxford University Press, 2005).

Baker, Chris and Phongpaichit, Pasuk, *A History of Thailand* (New York: Cambridge University Press, 2005).

Baradat, Leon P., *Political Ideologies: Their Origins and Impact*, 8th edition (Upper Saddle River, NJ: Prentice Hall, 2003).

Barnes, Trevor J. and Farish, Matthew, "Between Regions: Science, Militarism, and American Geography from World War to Cold War," *Annals of the Association of American Geographers* 96(2006): 807–826.

Barthes, Roland, *Mythologies*, translated by Annette Lavers (New York: Hill and Wang, 1972 [1957]).

Bauman, Zygmunt, *Modernity and the Holocaust* (Ithaca, NY: Cornell University Press, 1989).

Becker, Elizabeth, *When the War was Over: Cambodia and the Khmer Rouge Revolution* (New York: Public Affairs, 1998).

Bello, Walden, *Dilemmas of Domination: The Unmaking of the American Empire* (New York: Henry Holt and Company, 2005).

Belsey, Catherine, *Poststructuralism: A Very Short Introduction* (Oxford: Oxford University Press, 2002).

Bradley, Mark P., *Imagining Vietnam and America: The Making of Postcolonial Vietnam, 1919–1950* (Chapel Hill: The University of North Carolina Press, 2000).

Brown, Ian, *Cambodia* (Herndon,VA: Stylus Publishing, 2000).

Browning, Christopher R., *Nazi Policy, Jewish Workers, German Killers* (Cambridge: Cambridge University Press, 2000).

Buttimer, Anne and Seamon, David (eds), *The Human Experience of Space and Place* (London: Croom Helm, 1980).

Caldwell, John, "Demographers and the Study of Mortality: Scope, Perspectives, and Theory," *Annals of the New York Academy of Sciences* 954(2001): 19–34.

Carney, Timothy, "The Unexpected Victory," in *Cambodia 1975–1978: Rendezous with Death*, Karl D. Jackson (ed.) (Princeton: Princeton University Press, 1989)

Casey, Edward, *The Fate of Place: A Philosophical History* (Berkeley: University of California Press, 1998).

Chandler, David P., "Revising the Past in Democratic Kampuchea: When Was the Birthday of the Party?" *Pacific Affairs* 56(1983): 288–300.

—— "Seeing Red: Perceptions of Cambodian History in Democratic Kampuchea," in *Revolution and Its Aftermath in Kampuchea: Eight Essays*, David P. Chandler and Ben Kiernan (eds) (New Haven, CT: Yale University Southeast Asia Studies, 1983), 34 56.

—— *The Tragedy of Cambodian History: Politics, War, and Revolution Since 1945* (New Haven, CT: Yale University Press, 1991).

—— "From 'Cambodge' to 'Kampuchea': State and Revolution in Cambodia, 1863–1979," *Thesis Eleven* 50(1997): 35–49.

—— *Voices from S-21: Terror and History in Pol Pot's Secret Prison* (Berkeley: University of California Press, 1999).

—— *Brother Number One: A Political Biography of Pol Pot*, revised edition (Chiang Mai, Thailand: Silkworm Press, 2000).

—— *A History of Cambodia*, 3rd edition (Boulder, CO: Westview Press, 2000).

Chandler, David P., Kiernan, Ben, and Boua, Chanthou, (eds), *Pol Pot Plans the Future: Confidential Leadership Documents from Democratic Kampuchea, 1976–1977* (New Haven, CT: Yale University Southeast Asia Studies, 1988).

Church, Peter, *A Short History of South-East Asia*, revised edition (Singapore: John Wiley & Sons, 2003).

Clymer, Kenton, *Troubled Relations: The United States and Cambodia Since 1870* (Dekalb: Northern Illinois University Press, 2007).

Coe, Michael D., *Angkor and the Khmer Civilization* (New York: Thames & Hudson, 2003).

Colburn, Forrest D., *The Vogue of Revolution in Poor Countries* (Princeton: Princeton University Press, 1994).

Cosgrove, Denis and Domosh, Mona, "Author and Authority: Writing the New Cultural Geography," in *Place/Culture/Representation*, James Duncan and David Ley (eds) (New York: Routledge, 1993), 25–38.

Cox, Kevin, *Political Geography: Territory, State, and Society* (Malden, MA: Blackwell, 2002).

Crampton, Jeremy, "Maps as Social Constructions: Power, Communication, and Visualization," *Progress in Human Geography* 25(2001): 235–252.

Crang, Mike and Thrift, Nigel, "Introduction," in *Thinking Space*, Mike Crang and Nigel Thrift (eds) (New York: Routledge, 2000), 1–30.

Cresswell, Tim, *In Place/Out of Place: Geography, Ideology and Transgression* (Minneapolis: University of Minnesota Press, 1996).

—— "Falling Down: Resistance as Diagnostic," in *Entanglements of Power: Geographies of Domination/Resistance*, Joanne P. Sharp, Paul Routledge, Chris Philo, and Ronan Paddison (eds) (New York: Routledge, 2000), 256–268.

—— *Place: A Short Introduction* (Malden, MA: Blackwell, 2004).

Curtis, Bruce, "Foucault on Governmentality and Population: The Impossible Discovery," *Canadian Journal of Sociology* 27(2002): 505–533.

D'Amato, Paul, *The Meaning of Marxism* (Chicago, IL: Haymarket Books, 2006).

Dean, Mitchell, *Governmentality: Power and Rule in Modern Society* (Thousand Oaks, CA: Sage Publications, 1999).

Delaney, David, *Race, Place and the Law, 1836–1948* (Austin: University of Texas Press, 1998).

Del Casino, Jr., Vincent J. and Hanna, Stephen P., "Beyond the 'Binaries': A Methodological Intervention for Interrogating Maps as Representational Practices," *ACME: An International E-Journal for Critical Geographies* 4(2006): 34–56.

Deleuze, Gilles, *Foucault*, translated by S. Hand (Minneapolis: University of Minnesota Press, 1988).

Domosh, Mona, "Those 'Gorgeous Incongruities': Polite Politics and Public Space on the Streets of Nineteenth-Century New York City," *Annals of the Association of American Geographers* 88(1998): 209–226.

Doty, Roxanne L., *Imperial Encounters: The Politics of Representation in North-South Relations* (Minneapolis: University of Minnesota Press, 1996).

Driver, Felix, *Geography Militant: Cultures of Exploration and Empire* (Oxford: Blackwell, 2001).

Duiker, William J., *The Communist Road to Power in Vietnam*, 3rd edition (Boulder, CO: Westview Press, 1996).

—— *Ho Chi Minh: A Life* (New York: Theia, 2000).

Duncan, James and Ley, David (eds), *Place/Culture/Representation* (New York: Rougledge, 1993).

Duncan, James and Ley, David, "Introduction: Representing the Place of Culture," in *Place/Culture/Representation*, James Duncan and David Ley (eds) (New York: Routledge, 1993), 1–21.

Dunlop, Nic, *The Lost Executioner: A Journey to the Heart of the Killing Fields* (New York: Walker and Company, 2005).

Ebihara, May, "Revolution and Reformulation in Kampuchean Village Culture," in *The Cambodian Agony*, 2nd edition, David A. Ablin and Marlowe Hood (eds) (New York: M.E. Sharpe, 1990), 16–61.

—— "Memories of the Pol Pot Era in a Cambodian Village," in *Cambodia Emerges from the Past: Eight Essays*, Judy Ledgerwood (ed.) (DeKalb, IL: Northern Illinois University Press, 2002), 91–108.

Edkins, Jenny, *Poststructuralism and International Relations: Bringing the Political Back In* (Boulder, CO: Lynn Rienner Publishers, 1999).

Edelman, Murray, *Political Language: Words that Succeed and Policies that Fail* (New York: Academic Press, 1977).

Eisenstein, Zillah, *Sexual Decoys: Gender, Race and War* (New York: Zed Books, 2006).

Elbaum, Max, *Revolution in the Air: Sixties Radicals Turn to Lenin, Mao, and Che* (London: Verso, 2004).

Entrikin, Nicholas, *The Betweenness of Place: Towards a Geography of Modernity* (London: Macmillan, 1991).

Etcheson, Craig, *After the Killing Fields: Lessons from the Cambodian Genocide* (Lubbock: Texas Tech University Press, 2005).

Fanon, Frantz, *The Wretched of the Earth* (New York: The Grove Press, 1963).

—— *Black Skin, White Masks* (New York: The Grove Press, 1967).

Foucault, Michel, *The Archaeology of Knowledge and the Discourse on Language*, translated by A. Sheridan Smith (London: Pantheon Books, 1972).

—— *Discipline and Punish: The Birth of the Prison*, translated by A. Sheridan Smith (New York: Vintage Books, 1979).

—— "Questions on Geography," in *Power/Knowledge: Selected Interviews and Other Writings, 1972–1977*, C. Gordon (ed.) (New York: Pantheon Books, 1980), 63–77.

—— "Truth and Power," in *Power/Knowledge: Selected Interviews and Other Writings, 1972–1977*, C. Gordon (ed.) (New York: Pantheon Books, 1980), 109–133.

—— *The History of Sexuality, Volume I: An Introduction*, translated by R. Hurley (New York: Vintage Books, 1990).

—— "The Subject and Power," in *Power: Essential Works of Foucault, 1954–1984*, Vol. 3, J. Faubion (ed.) (New York: The New Press, 2000), 326–348.

—— "Technologies of the Self," in *Ethics: Subjectivity and Truth, Essential Works of Foucault, 1954–1984*, Vol. 1, P. Rabinow (ed.) (New York: The New Press, 2000), 223–251.

—— *Society Must Be Defended: Lectures at the Collège de France*, translated by D. Macey (New York: Picador, 2003).

Goodwin, Jeff, *No Other Way Out: States and Revolutionary Movements, 1945–1991* (Cambridge: Cambridge University Press, 2001).

Gottesman, Evan, *Cambodia After the Khmer Rouge: Inside the Politics of Nation Building* (New Haven, CT: Yale University Press, 2003).

Gregory, Derek, "Intervention in the Historical Geography of Modernity: Social Theory, Spatiality and the Politics of Representation," in *Place/Culture/Representation*, James Duncan and David Ley (eds) (New York: Routledge, 1993), 272–313.

—— *Geographical Imaginations* (Cambridge, MA: Blackwell, 1994).

—— "The Lightening of Possible Storms," *Antipode* 36(2004): 798–808.

Harley, J.B., "Maps, Knowledge, and Power," in *The Iconography of Landscape: Essays on the Symbolic Representation, Design and Use of Past Environments*, Denis Cosgrove and Stephen Daniels (eds) (Cambridge, UK: Cambridge University Press, 1988), 277–312.

—— "Deconstructing the Map," *Cartographica* 26(1989): 1–20.

—— "Cartography, Ethics, and Social Theory," *Cartographica* 27(1990): 1–23.

Harris, Ian M., and Morrison, Mary Lee, *Peace Education*, 2nd edition (London: McFarland Co., 2003).

Harvey, David, *The Limits to Capital* (Oxford: Basil Blackwell, 1982).

—— *The Condition of Postmodernity* (Oxford: Basil Blackwell, 1989).

—— *Justice, Nature and the Geography of Difference* (Oxford: Blackwell, 1996).

—— *Spaces of Hope* (Berkeley: University of California Press, 2000).

—— *Spaces of Capital: Towards a Critical Geography* (New York: Routledge, 2001).

Hearden, Patrick J., *The Tragedy of Vietnam: Causes and Consequences*, 2nd edition (New York: Pearson Longman, 2005).

Heder, Stephen and Tittemore, Brian D., *Seven Candidates for Prosecution: Accountability for the Crimes of the Khmer Rouge*, 2nd edition (Phnom Penh: Documentation Center of Cambodia, 2004).

Hinton, Alexander L., *Why Did They Kill? Cambodia in the Shadow of Genocide* (Berkeley: University of California Press, 2005).

Hossein-Zadeh, Ismael, *The Political Economy of US Militarism* (New York: Palgrave Macmillan, 2006).

Hubbard, Phil, Kitchin, Rob, Bartley, Brendan, and Fuller, Duncan, *Thinking Geographically: Space, Theory and Contemporary Human Geography* (New York: Continuum, 2002).

Hubbard, Phil, Kitchin, Rob, and Valentine, Gill (eds), *Key Thinkers on Space and Place* (Thousand Oaks, CA: Sage Publications, 2004).

Hynes, Patricia H., "On the Battlefield of Women's Bodies: An Overview of the Harm of War to Women," *Women's Studies International Forum* 27(2004): 431–445.

Ingersoll, David E., Matthews, Richard K., and Davison, Andrew, *The Philosophic Roots of Modern Ideology: Liberalism, Communism, Fascism, Islamism*, 3rd edition (Upper Saddle River, NJ: Prentice Hall, 2001).

Issacs, Arnold R., *Without Honor: Defeat in Cambodia* (Baltimore: Johns Hopkins University Press, 1983).

Jackson, Karl D., "The Ideology of Total Revolution," in *Cambodia 1975–1978: Rendezvous with Death*, Karl D. Jackson (ed.) (Princeton: Princeton University Press, 1989), 37–78.

Johnson, Chalmers, *The Sorrows of Empire: Militarism, Secrecy, and the End of the Republic* (New York: Metropolitan Books, 2004).

Johnston, Ron J., *Philosophy and Human Geography: An Introduction to Contemporary Approaches*, 2nd edition (London: Edward Arnold, 1986).

Kamm, Henry, *Report from a Stricken Land* (New York: Arcade Publishing, 1998).

Katz, Mark N., (ed.), *Revolution: International Dimensions* (Washington, DC: CQ Press, 2001).

Keen, David, *Endless War? Hidden Functions of the 'War on Terror'* (London: Pluto Press, 2006).

Kiernan, Ben, "Origins of Khmer Communism," *Southeast Asian Affairs* (1981): 161–180.

—— *How Pol Pot Came to Power: A History of Communism in Kampuchea, 1930–1975* (London: Verso, 1985).

—— *The Pol Pot Regime: Polices, Race and Genocide in Cambodia Under the Khmer Rouge, 1975–1979* (New Haven, CT: Yale University Press, 1996).

Knox, Paul, Agnew, John, and McCarthy, Linda, *The Geography of the World Economy*, 4th edition (London: Arnold, 2003).

Kuhlke, Olaf, *Representing German Identity in the New Berlin Republic: Body, Nation, and Place* (Lewiston: Edwin Mellen Press, 2005).

—— "Human Geography and Space," in *Encyclopedia of Human Geography*, Barney Warf (ed.) (Thousand Oaks, CA: Sage Publications, 2006), 441–444.

Kurlansky, Mark, *Nonviolence: Twenty-Five Lessons from the History of a Dangerous Idea* (New York: Modern Library, 2006).

Lefebvre, Henri, *The Production of Space*, translated by D. Nicholson-Smith (Oxford, UK: Blackwell, 1991 [1974]).

Legg, Stephen, "Foucault's Population Geographies: Classifications, Biopolitics and Governmental Spaces," *Population, Space and Place* 11(2005): 137–156.

Leitch, Vincent B., *Cultural Criticism, Literary Theory, Poststructuralism* (New York: Columbia University Press, 1992).

Lewis, Martin and Wigen, Kären, *The Myth of Continents: A Critique of Metageography* (Berkeley: University of California Press, 1997).

Locard, Henri, *Pol Pot's Little Red Book: The Sayings of Angkar* (Chiang Mai, Thailand: Silkworm Books, 2004).

Louw, Stephen, "In the Shadows of the Pharaohs: The Militarization of Labour Debate and Classical Marxist Theory," *Economy and Society* 29(2000): 239–263.

Mam, Kalyanee, "The Endurance of the Cambodian Family Under the Khmer Rouge Regime: An Oral History," in *Genocide in Cambodia and Rwanda: New Perspectives*, Susan E. Cook (ed.) (New Haven, CT: Yale Center for International and Area Studies, Genocide Studies Program Monograph Series No. 1, 2004), 127–171.

Marsden, Richard, "A Political Technology of the Body: How Labour is Organized into a Productive Force," *Critical Perspectives on Accounting* 9(1998): 99–136.

Marston, John, "Democratic Kampuchea and the Idea of Modernity," in *Cambodia Emerges from the Past: Eight Essays*, Judy Ledgerwood (ed.) (DeKalb, IL: Northern Illinois University Press, 2002), 38–59.

Massey, Doreen, *Space, Place, and Gender* (Minneapolis: University of Minnesota Press, 1994).

May, Todd, *The Philosophy of Foucault* (Montreal: McGill-Queen's University Press, 2006).

Mbembe, Achille, "Necropolitics," *Public Culture* 15(2003): 11–40.

McDowell, Linda, *Gender, Identity and Place: Understanding Feminist Geographies* (Minneapolis: University of Minnesota Press, 1999).

McIntyre, Kevin, "Geography as Destiny: Cities, Villages and Khmer Rouge Orientalism," *Comparative Studies in Society and History* 38(1996): 730–758.

McLaren, M.A. *Feminism, Foucault, and Embodied Subjectivity* (Albany: State University of New York Press, 2002).

McLellan, David (ed.), *Karl Marx: Selected Writings*, 2nd edition (Oxford: Oxford University Press, 2000).

Melson, Robert, *Revolution and Genocide: On the Origins of the Armenian Genocide and the Holocaust* (Chicago: University of Chicago Press, 1992).

Mitchell, Don, "The End of Public Space? People's Park, Definitions of the Public, and Democracy," *Annals of the Association of American Geographers* 85(1995): 108–133.

—— *The Right to the City: Social Justice and the Fight for Public Space* (New York: The Guilford Press, 2003).

Nagel, Caroline R., "Social Justice, Self-Interest and Salman Rushdie: Reassessing Identity Politics in Multicultural Britain," in *Geography and Ethics: Journeys in a Moral Terrain*, James D. Proctor and David M. Smith (eds) (New York: Routledge, 1999), 132–146.

Nast, Heidi and Pile, Steve (eds), *Places Through the Body* (London: Routledge, 1998).

Natter, Wolfgang and Jones III, John Paul, "Identity, Space, and Other Uncertainties," in *Space and Social Theory: Interpreting Modernity and Postmodernity*, Georges Benko and Ulf Strohmayer (eds) (Oxford, UK: Blackwell, 1997), 141–161.

Neale, Jonathan, *A People's History of the Vietnam War* (New York: The New Press, 2003).

Ngor, Haing (with R. Warner), *Survival in the Killing Fields* (New York: Carroll and Graf Publishers, 1987).

Nixon, Richard M., "Address to the Nation on the Situation in Southeast Asia," April 30, 1970, www.nixonlibrary.org.

Nowrojee, Binaifer, *Shattered Lives: Sexual Violence During the Rwandan Genocide and its Aftermath* (New York: Human Rights Watch, 1996).

Opotow, Susan, "Reconciliation in Times of Impunity: Challenges for Social Justice," *Social Justice Research* 14(2001): 149–170.

Opotow, Susan, Gerson, Janet, and Woodside, Sarah, "From Moral Exclusion to Moral Inclusion: Theory for Teaching Peace," *Theory into Practice* 44(2005): 303–318.

Orwell, George, *A Collection of Essays* (New York: Harcourt, 1981).

Ó Tuathail, Gearóid, *Critical Geopolitics: The Politics of Writing Global Space* (Minneapolis: University of Minnesota Press, 1996).

Paige, Jeffery M., *Agrarian Revolutions: Social Movements and Export Agriculture in the Underdeveloped World* (New York: The Free Press, 1975).

Parsa, Misagh, *States, Ideologies, and Social Revolutions: A Comparative Analysis of Iran, Nicaragua, and the Philippines* (Cambridge: Cambridge University Press, 2000).

Peet, Richard, *Global Capitalism: Theories of Societal Development* (New York: Routledge, 1991).

—— *Modern Geographical Thought* (Oxford, UK: Blackwell, 1998).

Philo, Chris, "Accumulating Populations: Bodies, Institutions and Space," *International Journal of Population Geography* 7(2001): 473–490.

—— "Sex, Life, Death, Geography: Fragmentary Remarks Inspired by 'Foucault's Population Geographies,'" *Population, Space and Place* 11(2005): 325–333.

Pitts, Victoria, *In the Flesh: The Cultural Politics of Body Modification* (New York: Palgrave Macmillan, 2003).

Ponchaud, François, "Social Change in the Vortex of Revolution," in *Cambodia 1975–1978: Rendezvous with Death*, Karl D. Jackson (ed.) (Princeton: Princeton University Press, 1989), 151–177.

Porter, Gareth, "Vietnamese Communist Policy Toward Kampuchea, 1930–1970," in *Revolution and Its Aftermath in Kampuchea: Eight Essays*, David P. Chandler and Ben Kiernan (eds) (New Haven, CT: Yale University Southeast Asia Studies, 1983), 57–98.

Porter, Philip W. and Sheppard, Eric S., *A World of Difference: Society, Nature, Development* (New York: The Guilford Press, 1998).

Pran, Dith (compiler) and DePaul, Kim (editor), *Children of Cambodia's Killing Fields: Memoirs by Survivors* (New Haven, CT: Yale University Press, 1997).

Quam, Louis O., "The Use of Maps in Propaganda," *Journal of Geography* 42(1943): 21–32.

Quinn, Kenneth M., "The Pattern and Scope of Violence," in *Cambodia 1975–1978: Rendezvous with Death*, Karl D. Jackson (ed.) (Princeton: Princeton University Press, 1989), 179–208.

Relph, Edward, *Place and Placelessness* (London: Pion, 1976).

Roberts, Paul C., "'War Communism': A Re-Examination," *Slavic Review* 29(1970): 238–261.

Sack, Robert D., *Conceptions of Space in Social Thought: A Geographic Perspective* (Minneapolis: University of Minnesota Press, 1980).

Said, Edward, *Orientalism* (New York: Vintage Books, 1979).

SarDesai, D.R., *Vietnam: Past and Present*, 4th edition (Cambridge, MA: Westview Press, 2005).

Sarup, Madan, *An Introductory Guide to Post-Structuralism and Postmodernism*, 2nd edition (Athens: University of Georgia Press, 1993).

Schulten, Susan, *The Geographical Imagination in America, 1880–1950* (Chicago: University of Chicago Press, 2001).

Schulzinger, Robert D., *A Time for War: The United States and Vietnam, 1941–1975* (New York: Oxford University Press, 1997).

Sedgwick, Eve, *Epistemology of the Closet* (Berkeley: University of California Press, 1990).

Seekins, Donald M., "Historical Setting," in *Cambodia: A Country Study*, R.R. Ross (ed.) (Washington, DC: US Government Printing Office, 1990), 3–71.

Selden, Raman and Widdowson, Peter, *A Reader's Guide to Contemporary Literary Theory*, 3rd edition (Lexington: University of Kentucky Press, 1993).

Shaw, John M., *The Cambodian Campaign: The 1970 Offensive and America's Vietnam War* (Lawrence: University of Kansas Press, 2005).

Shaw, Martin, *What is Genocide?* (Malden, MA: Polity, 2007).

Shawcross, William, *Sideshow: Kissinger, Nixon, and the Destruction of Cambodia*, revised edition (New York: Cooper Square Press, 2002).

Shields, Rob, *Places on the Margin* (London: Routledge, 1991).

—— "Spatial Stress and Resistance: Social Meanings of Spatialization," in *Space and Social Theory: Interpreting Modernity and Postmodernity*, Georges Benko and Ulf Strohmayer (eds) (Malden, MA: Blackwell, 1997), 186–202.

—— *Lefebvre, Love and Struggle: Spatial Dialectics* (New York: Routledge, 1999).

Short, Philip, *Pol Pot: Anatomy of a Nightmare* (New York: Henry Holt and Company, 2004).

Singer, Peter, *Marx: A Very Short Introduction* (Oxford, UK: Oxford University Press, 2000).

Smith, David, *Geography and Social Justice* (Cambridge, MA: Blackwell, 1994).

Soja, Edward W., *Postmodern Geographies: The Reassertion of Space in Critical Social Theory* (New York: Verso, 1989).

Somekawa, Ellen and Smith, Elizabeth A., "Theorizing the Writing of History or, 'I Can't Think Why it Should be so Dull, for a Great Deal of it Must be Invention,'" *Journal of Social History* 22(1998): 149–162.

Sparke, Matthew, "A Map that Roared and an Original Atlas: Canada, Cartography, and the Narration of Nation," *Annals of the Association of American Geographes* 88(1998): 463–475.

Stewart, Lynn, "Bodies, Visions, and Spatial Politics: A Review Essay on Henri Lefebvre's *The Production of Space*," *Environment and Planning D: Society and Space* 13(1995): 609–618.

Strub, H., "The Theory of Panoptical Control: Bentham's Panopticon and Orwell's *Nineteen Eighty-Four*," *The Journal of the History of the Behavioral Sciences* 25(1989): 40–49.

Tarrow, Sidney, *Power in Movement: Social Movements and Contentious Politics*, 2nd edition (Cambridge: Cambridge University Press, 1998).

Taylor, Peter J., "God Invented War to Teach Americans Geography," *Political Geography* 23(2004): 487–492.

Thion, Serge, "The Cambodian Idea of Revolution," in *Revolution and Its Aftermath in Kampuchea: Eight Essays*, David P. Chandler and Ben Kiernan (eds) (New Haven, CT: Yale University Southeast Asia Studies, 1983), 10–33.

Thomas, Louis B., "Maps as Instruments of Propaganda," *Surveying and Mapping* 9(1949): 75–81.

Thrower, Norman J.W., *Maps and Civilization: Cartography in Culture and Society* (Chicago: University of Chicago Press, 1996).

Tuan, Yi-Fu, *Topophilia: A Study of Environmental Perception, Attitudes and Values* (Englewood Cliffs, NJ: Prentice Hall, 1974).

—— *Space and Place: The Perspective of Experience* (Minneapolis: University of Minnesota Press, 1977).

Tully, John, *A Short History of Cambodia: From Empire to Survival* (Crows Nest, Australia: Allen & Unwin, 2005).

Tyner, James A. *The Geography of Malcolm X: Black Radicalism and the Remaking of American Space* (New York: Routledge, 2006).

—— *The Business of War: Workers, Warriors and Hostages in Occupied Iraq* (Aldershot, UK: Ashgate, 2006).

—— *America's Strategy in Southeast Asia: From the Cold War to the Terror War* (Boulder, CO: Rowman & Littlefield, 2007).

Tyner (née Zink), Judith A., *Persuasive Cartography: An Examination of the Map as a Subjective Tool of Communication*, PhD Dissertation, University of California, Los Angeles, 1974.

—— "Persuasive Cartography," *Journal of Geography* 81(1982): 140–144.

—— "Images of the Southwest in Nineteenth-Century American Atlases," in *The Mapping of the American Southwest*, Dennis Reinhartz and Charles C. Colley (eds) (College Station: Texas A&M University Press, 1987), 57–77.

Valentino, Benjamin A., *Final Solutions: Mass Killing and Genocide in the 20th Century* (Ithaca, NY: Cornell University Press, 2004).

Vickery, Michael, "Democratic Kampuchea—Themes and Variations," in *Revolution and its Aftermath in Kampuchea: Eight Essays*, David P. Chandler and Ben

Kiernan (eds) (New Haven, CT: Yale University Southeast Asia Studies, 1983), 99 136.

—— *Society, Economics, and Politics in Pre-Angkor Cambodia* (Tokyo: Center for East Asian Cultural Studies for UNESCO, 1998).

Vriens, Lennart, "Peace Education: Cooperative Building of a Humane Future," *Pastoral Care in Education* 15(1997): 25–30.

Waller, James, *Becoming Evil: How Ordinary People Commit Genocide and Mass Killing* (Oxford: Oxford University Press, 2002).

Weedon, Chris, *Feminist Practice and Poststructuralist Theory*, 2nd edition (Malden, MA: Blackwell Publishers, 1997).

Weeks, John R., *Population: An Introduction to Concepts and Issues*, 9th edition (Belmont, CA: Wadsworth/Thomsom Learning, 2005).

Wood, Denis, *The Power of Maps* (New York: The Guilford Press, 1992).

Wood, Ellen M., *Empire of Capital* (London: Verso, 2003).

Wyatt, David K., *Thailand: A Short History* (New Haven, CT: Yale University Press, 1984).

Young, Iris M., *Justice and the Politics of Difference* (Princeton, NJ: Princeton University Press, 1990).

—— "Harvey's Complaint with Race and Gender Issues: A Critical Response," *Antipode* 30(1998): 36–42.

Yuval-Davis, Nira, "Women and the Biological Reproduction of 'the Nation,'" *Women's Studies International Forum* 19(1996): 17–24.

Zinn, Howard, *Just War* (Milan, Italy: Edizioni Charta, 2005).

Index